THE NATIONAL ACADEMIES
Advisers to the Nation on Science, Engineering, and Medicine

The **National Academy of Sciences** is a private, nonprofit, self-perpetuating society of distinguished scholars engaged in scientific and engineering research, dedicated to the furtherance of science and technology and to their use for the general welfare. Upon the authority of the charter granted to it by the Congress in 1863, the Academy has a mandate that requires it to advise the federal government on scientific and technical matters. Dr. Ralph J. Cicerone is president of the National Academy of Sciences.

The **National Academy of Engineering** was established in 1964, under the charter of the National Academy of Sciences, as a parallel organization of outstanding engineers. It is autonomous in its administration and in the selection of its members, sharing with the National Academy of Sciences the responsibility for advising the federal government. The National Academy of Engineering also sponsors engineering programs aimed at meeting national needs, encourages education and research, and recognizes the superior achievements of engineers. Dr. Charles M. Vest is president of the National Academy of Engineering.

The **Institute of Medicine** was established in 1970 by the National Academy of Sciences to secure the services of eminent members of appropriate professions in the examination of policy matters pertaining to the health of the public. The Institute acts under the responsibility given to the National Academy of Sciences by its congressional charter to be an adviser to the federal government and, upon its own initiative, to identify issues of medical care, research, and education. Dr. Harvey V. Fineberg is president of the Institute of Medicine.

The **National Research Council** was organized by the National Academy of Sciences in 1916 to associate the broad community of science and technology with the Academy's purposes of furthering knowledge and advising the federal government. Functioning in accordance with general policies determined by the Academy, the Council has become the principal operating agency of both the National Academy of Sciences and the National Academy of Engineering in providing services to the government, the public, and the scientific and engineering communities. The Council is administered jointly by both Academies and the Institute of Medicine. Dr. Ralph J. Cicerone and Dr. Charles M. Vest are chair and vice chair, respectively, of the National Research Council.

www.national-academies.org

Laura D'Andrea Tyson, S. K. and Angela Chan Professor of Global Management, Haas School of Business, University of California at Berkeley
Hal Varian, Chief Economist, Google, Inc.
Charles M. Vest (ex-officio), President, National Academy of Engineering
Alan Wm. Wolff, Senior Counsel, McKenna, Long & Aldridge LLP

Staff

Stephen A. Merrill, Executive Director
Charles W. Wessner, Program Director
Paul T. Beaton, Program Officer
McAlister Clabaugh, Program Officer
Aqila Coulthurst, Program Coordinator
David Dawson, Senior Program Assistant
Sujai Shivakumar, Senior Program Officer
David Dierksheide, Program Officer
Cynthia Getner, Financial Officer

Preface

The U.S. Congress directed the U.S. Department of the Treasury to request that the National Academy of Sciences undertake "a comprehensive review of the Internal Revenue Code of 1986 to identify the types of and specific tax provisions that have the largest effects on carbon and other greenhouse gas emissions and to estimate the magnitude of those effects" (P.L. 110-343, Division B, Title I, Sec. 117). Congress appropriated funds for this study in its 2010 appropriations (P.L. 111-117; Division C, Title I, Sec. 126).

After the National Academies accepted this assignment, the National Research Council established the ad hoc Committee on the Effects of Provisions in the Internal Revenue Code on Greenhouse Gas Emissions, which prepared this report. Appendix B contains biographical information on the committee members.

The committee met five times as it worked to prepare this report. At its first meeting in April 2011, the committee held an open session for interested members of the public to make presentations to the committee. The following individuals responded to notice of that open session and made oral presentations to the committee: Elizabeth Paranhos (on behalf of Environmental Defense Fund); Jay Pendergrass (Environmental Law Institute); and Eric Pica (Friends of the Earth). Two individuals could not attend in person but submitted written statements: Janet Milne (Environmental Tax Policy Institute at Vermont Law School); and Douglas Koplow (Earth Track). Additionally, the following individuals made presentations at the invitation of the committee: Mun Ho (Resources for the Future); Gilbert Metcalf (Tufts University); and Ian Parry (International Monetary Fund). During later meetings, the committee also requested presentations from the following individuals: Alan Krupnik (Resources for the Future); Stephen P. A. Brown (University of Nevada, Las Vegas); and Lessly Goudarzi and Frances Wood (OnLocation, Inc.). We are grateful for the thoughtful presentations that these individuals made.

The committee also made use of peer-reviewed scientific literature, working papers, government agency reports, and think tank reports as it deliberated and in producing this report. The committee extends its thanks to Danny Cullenward (Stanford University), Kathleen Foreman (University of California,

Berkeley), and Ritadhi Chakravarti (formerly of the Urban Institute) for help with reviews of relevant literature. These literature reviews were crucial for understanding and framing the need to undertake original economic modeling in order to respond to Congress's request.

At the core of this study is the committee's analysis of original economic modeling of tax policies performed by independent consultants. This report would not have been possible without their expertise and willingness to work with the committee through an iterative and often challenging process to understand each model's capabilities and, thus, which tax policies it could realistically model. All of the consultants produced excellent reports explaining results for the committee and did so in a timely manner and responsive to the specifications outlined by the committee. Readers can download those reports at the National Academies Press website, http://www.nap.edu/catalog.php?record_id=18299. The committee and the nation are indebted to the following: Lessly Goudarzi (OnLocation, Inc.), Frances Wood (OnLocation, Inc.), Dale W. Jorgenson (Dale Jorgenson Associates), Richard Goettle (Dale Jorgenson Associates), Wyatt Thompson (The Food and Agriculture Policy Research Institute, University of Missouri), Stephen P. A. Brown (Center for Business and Economic Research, Lee Business School, University of Nevada, Las Vegas), and Ryan Kennelly (Center for Business and Economic Research, Lee Business School, University of Nevada, Las Vegas).

Before committing funds, staff time, and other resources to the modeling exercises, the committee asked for independent experts to consider the choices of models to use and underlying methodology and offer critiques and suggestions. The committee is grateful to Richard Newell (Nicholas School of the Environment, Duke University) and William Pizer (Sanford School of Public Policy, Duke University) for carefully evaluating its plan for modeling selected tax policies and the committee's deliberative process for choosing those policies. Their comments and suggestions proved helpful in shaping the committee's final requests to the independent modeling consultants. We are highly indebted for their assistance. Despite this assistance, the committee remains solely responsible for all modeling decisions.

This report has been reviewed in draft form by individuals chosen for their diverse perspectives and technical expertise, in accordance with procedures approved by the National Academies' Report Review Committee. The purpose of this independent review is to provide candid and critical comments that will assist the institution in making its published report as sound as possible and to ensure that the report meets institutional standards for objectivity, evidence, and responsiveness to the study charge. The review comments and draft manuscript remain confidential to protect the integrity of the process.

We wish to thank the following individuals for their review of this report: Amos Avidan, Bechtel Corporation; John Birge, University of Chicago; Steven Davis, University of California, Irvine; Austen Goolsbee, University of Chicago; Lawrence Goulder, Stanford University; Russell Lee, Oak Ridge National Laboratory; Bruce McCarl, Texas A&M University; Dave McCurdy, American Gas Association; Gilbert Metcalf, Tufts University; Peter Merrill, PricewaterhouseCoopers LLP; Janet Milne, University of Vermont; William Pizer, Duke University; Mark Schwartz, PIRA University; Philip Tabas, The Nature Conservancy; Susan Tierney, Analysis Group Inc.; David Weisbach, University of Chicago; and John Weyant, Stanford University.

Although the reviewers listed above have provided many constructive comments and suggestions, they were not asked to endorse the conclusions or recommendations, nor did they see the final draft of the report before its release. The review of this report was overseen by T.J. Glauthier, TJG Energy Associates, LLC and Charles Manski, Northwestern University. Appointed by the National Academies, they were responsible for making certain that an independent examination of this report was carried out in accordance with institutional procedures and that all review comments were carefully considered. Responsibility for the final content of this report rests entirely with the authoring committee and the institution.

The committee could not have completed its work without the assistance of the talented and dedicated staff of the National Research Council's Board on Science, Technology, and Economic Policy (STEP). Stephen Merrill (director of STEP) served as study director, assisted by Paul Beaton (STEP) and Aqila Coulthurst (STEP). The committee is also indebted to Lint Barrage (Yale University), who served as an independent consultant to the committee, providing assistance with literature reviews, technical advice, and regularly briefing the committee on the details of relevant topics when requested. Several participants of the National Academies' Christine Mirzayan Science and Technology Policy Fellowship program provided additional scientific and technical assistance: Christopher Avery, Carrie Chen, Adnan Aslam, Marilyn Waite, and Vincent Huang. Karin Matchett served as an editor of the final report.

William D. Nordhaus, *Chair*
Committee on Effects of Provisions
in the Internal Revenue Code on
Greenhouse Gas Emissions

Contents

SUMMARY 1

1   OVERVIEW AND SCOPE OF THE STUDY 11

2   METHODS FOR EVALUATING TAX POLICY EFFECTS
ON GREENHOUSE GAS EMISSIONS...23

3   ENERGY-RELATED TAX EXPENDITURES...53

4   ENERGY-RELATED EXCISE TAXES 81

5   BIOFUELS SUBSIDIES...91

6   GREENHOUSE GAS EMISSIONS AND BROAD-BASED
TAX EXPENDITURES 113

7   SUMMARY OF FINDINGS AND RECOMMENDATIONS
AND USE OF TAX POLICY TO ADDRESS CLIMATE
CHANGE POLICY...135

REFERENCES...157

APPENDIXES

A   MODELING APPROACHES TO THE EFFECTS OF TAX
POLICY ON GHG EMISSIONS 165

B   BIOGRAPHICAL INFORMATION OF COMMITTEE
AND STAFF 175

TABLES AND FIGURES

TABLES

2-1   The 10 Largest Excise Tax Collections for Fiscal Year 2010, 28

2-2   The 10 Largest Energy Tax Policies (by dollars of foregone revenue), 30
2-3   The 10 Largest Broad-based Tax Expenditures, 31
2-4   Provisions Modeled for This Study and Where Discussed in
       This Report, 49
3-1   Assumptions Underlying NEMS Scenarios, 56
3-2a   Summary of CO_2 Emissions Impacts, 59
3-2b   Summary of CO_2 Emissions Impacts, 60
4-1   Summary Impacts of Removing Federal Highway Fuels Taxes
       Across Four Models, 83
4-2   Summary Appraisal of Studies of Impact of Removing Highway
       Fuels Taxes, 88
5-1   Key Modeling Assumptions, 97
5-2   Removal of Biofuel Provisions – Key Modeling Results, 100
5-3   Effect of the RFS2 Mandate on Model Projections for the "Removing
       All Biofuel Provisions" Scenario: Key Modeling Results, 105
5-4   Alternative Emission Coefficients for Gasoline and Biofuels Based
       on Study, 107
5-5   Sensitivity of GHG Impacts from Variations in Biofuel GHG Emission
       Coefficients: Removing all Provisions Scenario, 107
6-1   Energy Intensities of Different Sectors, 116
6-2   Effects of Different Revenue Recycling Options, 118
6-3   Estimated Effects Relative to Base Case of Eliminating Accelerated
       Depreciation Preference on Key Economic Variables, 2010-2035, 118
6-4   Estimated Effects of Eliminating the Home Mortgage Interest Deduction
       Relative to Base Case, 2010-2035, 127
6-5   Estimated Effects Relative to Base Case of Eliminating the Exclusion for
       Employer-supplied Health Insurance, 2010-2035, 132
7-1   Summary of Modeling Results, 150

FIGURES

1-1   Schematic representation of how taxes affect GHG emissions
       through the market, 18
3-1   Total U. S. Electricity Generation, in Terawatt hours (TWh) – Reference
       Scenario and No-PTC/ITC Scenario, 64
3-2   Changes in Electricity Generation Caused, in Terawatt hours (TWh), by
       Removing the PTC/ITC Compared to the Reference Scenario, 65
3-3   Changes in Electricity Generating Capacity Caused by Removing
       the PTC/ITC Compared to the Reference Scenario, 65
3-4   Changes in Non-Hydro Renewable Generation Caused by Removing
       the PTC/ITC Compared to the Reference Scenario, 66
3-5   U.S. Average Retail Electricity Prices under the Reference Scenario
       and the No-PTC/ITC Scenario, 66
3-6   PTC/ITC Tax Expenditures – Various Scenarios, 68
3-7   Power Sector CO_2 Emissions – Four Scenarios, 69
3-8   Natural Gas Consumption Under Three Scenarios, 72

3-9   Natural Gas Prices Under Several Scenarios, 73
3-10   Changes in CO_2 Emissions under the Cost Depletion Scenario, 75
5-1   Effects of biofuel provisions removal on biofuel production levels, 108
5-2   Effects of removing the biofuel provisions on total energy
        expenditures, 109
5-3   Effects of biofuel provisions removal on federal revenue, 109
5-4   Effects of biofuel provisions removal on U.S. CO_2 emissions, 109

Summary

STUDY ORIGINS AND SCOPE

The U.S. Congress charged the National Academies with conducting "a comprehensive review of the Internal Revenue Code to identify the types of and specific tax provisions that have the largest effects on carbon and other greenhouse gas emissions and to estimate the magnitude of those effects." To address such a broad charge, the National Academies appointed a committee composed of experts in tax policy, energy and environmental modeling, economics, environmental law, climate science, and related areas.

For scientific background informing the study, the committee relied on the earlier findings and studies by the National Academies, the U.S. government, and other research organizations. The committee has relied on earlier reports and studies to set the boundaries of the economic, environmental, and regulatory assumptions for the present study. The major economic and environmental assumptions are those developed by the U.S. Energy Information Administration (EIA) in its annual reports and modeling. Additionally, the committee has relied upon publicly available data provided by the U.S. Environmental Protection Agency, which inventories greenhouse gas (GHG) emissions from different sources in the United States.

The tax system affects emissions primarily through changes in the prices of inputs and outputs or goods and services. Most of the tax provisions considered in this report relate directly to the production or consumption of different energy sources. However, there is a substantial set of tax expenditures that we call "broad-based" that favor certain categories of consumption—among them, employer-provided health care, owner-occupied housing, and purchase of new plants and equipment. The committee examined both tax expenditures and excise taxes that could have a significant impact on GHG emissions.

SELECTION OF TAX PROVISIONS AND METHODS OF ANALYSIS

Limited time and other resources compelled the committee to focus its work. Accordingly, the committee decided to concentrate its attention on four groups of tax code provisions and some related regulatory policies. Table 2-4 (Chapter 2) lists the tax code provisions examined and their associated revenue

1

consequences and lists the chapters where the analysis of each provision is discussed.

In the end, the committee analyzed tax provisions that account for 46 percent of all energy-related excise tax revenues as well as those accounting for 71 percent of the calculated revenue loss from the 10 largest energy-related tax expenditures in 2011. As estimated by the Treasury Department, the broad-based tax expenditures selected account for about one-third of the cost of all tax expenditures that year.

REVIEW OF EXISTING RESEARCH

The next step was to review existing research on the impact of the tax code on greenhouse gas emissions. This was undertaken by the committee, the staff, and a team of consultants hired specifically for this study.

The committee found that a substantial body of research relating tax policy to greenhouse gas emissions is limited to two areas: highway taxes and emissions taxes. Studies of the impact of highway motor fuels taxes, particularly those on gasoline, go back decades. However, most of these studies do not incorporate important features of the U.S. tax or other regulatory mandates, such as biofuels taxes and subsidies, or Corporate Average Fuel Economy (CAFE) standards, and very few examine the impacts on GHG emissions. Moreover, most studies do not incorporate the GHG impacts coming from linkages to the rest of the economy.

A second set of studies examines the efficiency and effectiveness of taxes on GHG emissions. These studies incorporate both empirical studies as well as model simulations. Such studies include both individual models and model comparison studies, and consider taxes for the United States as well as for other countries.

In view of the insufficiency of existing research and need to use a unified set of baseline assumptions to compare effects of different tax provisions, the committee concluded that it would be necessary to commission new economic modeling studies capable of estimating these effects on investment decisions, their effects in turn on energy production and consumption, and the resulting effects on emissions. Chapter 2 describes in detail the models and the rationale for their selection.

The models used in the analysis have different structures and assumptions (as described in Appendix A), and most are limited in their capacity. For example, some were unable to analyze global as well as U.S. emissions, and only one could analyze the general equilibrium or economy-wide impacts of tax policies. For these reasons, readers should regard the numerical results as suggestive rather than definitive.

For each of the models, the committee specified a set of baseline assumptions on gross domestic product (GDP) growth, oil prices, and the regulatory environment, as well as the tax system. The rate of U.S. GDP growth and path of world oil prices are those used in the Energy Information Administration's baseline, the Annual Energy Outlook for 2011 (AEO11), used widely in the energy and economic modeling community. The tax code and regulatory environment of 2011 was chosen by the committee as the basis of its analysis, in part because these are also included in the AEO11 baseline. Tax code provisions that have expired or are scheduled to expire are assumed to be extended indefinitely in our reference scenario. The committee took regulations in place in 2011 as the regulatory baseline. The time period of the study was generally 2010–2035. To estimate the effects on emissions of particular tax code provisions, the committee instructed the modeling consultants to run scenarios where they removed each of the taxes or tax preferences from the baseline one at a time, keeping all other policies, assumptions, and taxes unchanged. One model, which focused on agricultural markets, is designed to represent changes in global GHG emissions. The others—two focused on the energy sector and one focused economy-wide— can estimate only domestic U.S. emissions. Nevertheless, the first-order tax policy effects of principal interest to Congress are on U.S. GHG emissions.[1]

PROVISION-BY-PROVISION FINDINGS

Production Tax Credits for Renewable Electricity

The production and investment tax credits for renewable electricity provide a tax credit of 2.3 cents per kWh of power for the first 10 years of electricity production generated from qualifying renewable sources (primarily solar, wind, and biomass) or a credit equal to 30 percent of investment in qualifying equipment. These credits lower the cost of electricity generated from renewable resources, encouraging their substitution for fossil fuels and thereby tend to reduce GHG emissions. The committee's analysis indicates that these provisions do lower CO_2 emissions under the macroeconomic conditions in the AEO11 reference and high GDP growth cases, but the impact is small, about 0.3 percent of U.S. CO_2 in the reference case.

Oil and Gas Depletion Allowances

The percentage depletion allowance permits independent (nonintegrated) domestic producers of oil and gas to deduct a percentage of gross income associated with sale of the commodity up to certain limits. The depletion rate is set at 15 percent of gross revenues associated with production. In modeling completed

---

[1]This original modeling was undertaken by four independent consultants. Each of those consultants produced reports to the committee detailing the results of their modeling efforts. Readers can download those reports at the National Academies Press website, http://www.nap.edu/catalog.php?record_id=18299.

for this report, removing the percentage depletion allowances (and substituting cost depletion) has virtually no effect on oil production and associated GHG emissions. Although natural gas production goes down as the tax preference is removed, the complex substitution patterns among fuels lead to offsetting market forces and to a minimal impact on overall emissions.

Home Energy-efficient Improvement Credits

The committee examined Credits for Energy Efficiency Improvements to Existing Homes qualitatively, because time and budget constraints precluded obtaining detailed and reliable estimates of its impacts. Using market analysis, the committee expects that this credit is unlikely to produce major reductions in GHG emissions. However, the size of the tax expenditure and the evidence of unexploited energy-efficiency gains in the housing sector led the committee to conclude that this provision merits high priority for future research.

Nuclear Decommissioning Tax Preference

A further provision that was analyzed qualitatively was the special tax rate on reserves set up to decommission nuclear power plants at the end of their lifetime. Based on the available evidence, including the projections of nuclear power under different scenarios, the committee finds that the decommissioning provision is likely to have little impact on GHG emissions.

Highway Motor Fuels Taxes

The federal excise taxes on highway motor fuels in 2011 included a tax of $0.183 per gallon for gasoline, $0.243 per gallon for diesel fuel and kerosene, and $0.197 per gallon for diesel-water fuel emulsion. This chapter reviewed four commissioned studies of the effect of removing the excise taxes on highway fuels. All four models find that removing the excise taxes on highway fuels would result in increasing greenhouse gas emissions. But the magnitude of the estimated effects varies dramatically for the different models.

Having studied the model results and the broader literature, the committee concludes that the differences among the models are large and incompletely understood. The differences arise from the types and values of price elasticities used by the different models, from assumptions about increasing biofuels production and consumption to meet the RFS mandates, from the volumetric bias of highway fuels taxes, and from application of the tax within each model's structure. A close examination of the results leads the committee to conclude that the NEMS-NAS and the FAPRI models capture the forces at work in this sector most reliably and therefore form the basis of our estimates. Taking these two modeling results together produces a striking conclusion: The impact of removing highway fuels taxes on GHG emissions is estimated to be very small because of special features of the taxes and the market. The results on highway taxes are contingent on special features of this market because the results depend upon the structure, timing, and implementation of the renewable fuels standards (RFS) as well as a quirk in the tax structure (its volumetric bias). The magnitude of the differences across models leads the committee to caution against relying on specific numerical results from a single model and recommends drawing only broad conclusions about the nature and direction of impacts. Policy makers and analysts should rely on multiple models, methodologies, and estimates in calculating impact of the tax code and other policies on greenhouse-gas emissions and climate change.

Aviation Fuel Taxes

The federal excise tax for commercial aviation fuel is $0.043 per gallon and $0.193 for noncommercial aviation. The exhaustive literature searches did not find any detailed study of the impact of these taxes on GHG emissions. Additionally, the models used for detailed analysis in this study were unable to adequately represent the taxes. This, therefore, is a high-priority area for further work given that aviation is producing rapidly growing emissions, and as yet there are no substitutes for jet fuels.

Biofuels Provisions

One particularly important set of tax provisions involves the use of ethanol and other biofuels, particularly as substitutes for petroleum products. These provisions involve a complex combination of taxes, tax expenditures, import tariffs, and regulatory mandates that interact to change the composition of fuels and even affect agriculture. Most of these provisions expired in 2012, but under the committee's methodology, each of these provisions is included in the reference scenario.

The committee analyzed the biofuels provisions with three different models, although it concentrated its analysis on the Food and Agricultural Policy Research Institute at the University of Missouri (FAPRI-MU) model, which had the most detailed treatment of the biofuels sector. The findings indicate that removing all tax code provisions and the import tariff would result in a decrease of emissions of 5 million metric tons (MMT) per year of CO_2 equivalent globally. This is less than 0.02 percent of global emissions. The results are complicated by the mandates for renewable fuels. If the mandates are removed along with the subsidies, the estimated emissions are smaller than the estimates with the mandates. The results of the other modeling studies are consistent with the central FAPRI estimates.

These results show the often-counterintuitive nature of the effects of tax subsidies. Although it may seem obvious that subsidizing biofuels should reduce

CO_2 emissions because they rely on renewable resources rather than fossil fuels, many studies we reviewed found the opposite. As structured, the biofuels tax credits encouraged the consumption of motor fuels because they lower prices, and this effect appears to offset any reduction in the GHG intensity of motor fuels that occurs because of the incentives to blend biofuels with gasoline.

Accelerated Depreciation

Accelerated depreciation is one of the largest business tax expenditures in the federal income tax code. This set of provisions allows businesses to write off the value of their capital assets at a rate that is faster than the estimated economic depreciation. Modeling runs indicate that eliminating accelerated depreciation would reduce the GHG intensity of national output by shifting production away from GHG-intensive activities such as coal mining and electric power generation to low-GHG activities such as communications. However, the net effect depends upon how the resulting revenues are recycled. If the revenues were to be returned by lowering marginal tax rates, the net impact on GHG emissions is expected to be negligible. If revenues are refunded through lump-sum rebates, however, then GHG emissions should decrease.

Owner-occupied Housing Provisions

The significant incentives in the federal income tax code for owner-occupied housing include deductibility of mortgage interest and property taxes and exclusion from taxation of the first $250,000 ($500,000 for couples) of capital gains on home sales. The estimates prepared for the committee suggest that eliminating the tax subsidies for owner-occupied housing and using the revenue to lower marginal tax rates would improve the efficiency of allocation of the capital stock and increase national output. GHG emissions would increase at about the same rate as GDP increases. However, the simulation does not fully capture the effects of the subsidies on housing size or materials (affecting energy consumption) or location (changing patterns of automobile use and gasoline consumption). We therefore find the results inconclusive, underscoring the need for models that integrate effects on the housing stock with general equilibrium effects.

Employer-provided Health Care Provisions

The exclusion of employer-provided health insurance from the taxable income of employees is the largest single tax expenditure in the Internal Revenue Code. The committee expected that eliminating health care subsidies would raise GHG emissions per unit of output because the health care sector is less GHG intensive than the rest of the economy. The Intertemporal General Equilibrium Model (IGEM) results show the opposite effect, however, with a small decrease in GHG intensity. The committee's inability to understand the structural features of the model that produced these results leads it to conclude that the impact of the health provisions on GHG emissions remains an open question and an important subject for future research.

Further Observations on Broad-based Tax Provisions

The committee's major finding is that the broad-based provisions influence GHG emissions primarily through their effects on overall national output. In most cases, the percentage change in GHG emissions was close to or equal to the percentage change in national output induced by removing the tax provision. A second finding is that the way the revenues generated by eliminating tax preferences are recycled significantly affects output and emissions. A third finding is that the broad-based provisions generally have little effect on emissions intensities. Finally, the committee reiterates that the results are highly sensitive to assumptions about how tax revenues from eliminating the provisions are returned to the economy. We conclude that changes in broad-based tax provisions are likely to have a small impact on overall GHG emissions except through the impact on economic output. However, we caution that these results rely on a single model and therefore require further study.

COMPARISON WITH CBER MODELING RESULTS

We compared the results of our detailed modeling with those of a comprehensive study of energy tax expenditures by a modeling group at the University of Nevada at Las Vegas's Center for Business and Economic Research (CBER). The committee used the CBER model to obtain an order-of-magnitude estimate of the impact of all energy-related tax expenditures. Under the methods and assumptions of that study, if all tax subsidies would have been removed, then net CO_2 emissions would have decreased by 30 MMT per year over the 2005-2009 period. This total represented about ½ percent of total U.S. CO_2 emissions over this period. The CBER results are consistent with the basic findings of the detailed modeling studies we conducted—that the overall effect of current energy tax subsidies on GHG emissions is close to zero.

GENERAL FINDINGS

Our report does not estimate an aggregate impact of tax provisions on greenhouse gas emissions due to the complexity of the tax code as well as the difficulty of determining the impact of several important provisions. The summary table of impacts of different studies and provisions is contained in Table 7-1. The following provides a summary of the results from different sectors.

First, the combined effect of current energy-sector tax expenditures on GHG emissions is very small and could be negative or positive. The most comprehensive study available suggests that their combined impact is less than 1 percent of total U.S. emissions. If we consider the estimates of the effects of the provisions we analyzed using more robust models, they are in the same range. We cannot say with confidence whether the overall effect of energy-sector tax expenditures is to reduce or increase GHG emissions.

Second, individual energy-sector tax expenditures in some cases contribute to, and in other cases subtract from, U.S. and global GHG emissions. The subsidies on ethanol that expired in 2012 clearly added to global GHG emissions. By contrast, the balance of the evidence is that the production and investment tax credits for renewable electricity slightly reduce U.S. GHG emissions. The depletion allowance has virtually zero impact on emissions.

Third, the best existing analytical tools are unable to determine in a reliable fashion the impact of some important subsidies. Important tax expenditures that have resisted analysis include ones subsidizing residential energy efficiency. The difficulties in this case involve such factors as the discount rate consumers apply to future fuel savings, the strength of any rebound effect, and the extent to which consumers understand and respond to tax law changes.

Fourth, the revenues foregone by energy-sector tax subsidies are substantial in relation to the effects on GHG emissions. The Treasury estimates that the revenue loss from energy-sector tax expenditures in fiscal years 2011 and 2012 totaled $48 billion. Few of these were enacted to reduce GHG emissions. As policies to reduce GHG emissions, however, they are inefficient. Very little if any GHG reductions are achieved at substantial cost with these provisions.

Fifth, the emissions impacts of the broad-based tax expenditures are primarily through their impact on the level of national output. Broad-based tax expenditures entail roughly 50 times more revenues foregone than the energy-sector subsidies. We investigated a subset of provisions representing about one-third of the revenue losses from tax expenditures—subsidies to equipment investment through accelerated depreciation, to health care, and to owner-occupied housing. Except for accelerated depreciation, we were unable to reach a definite conclusion on whether they increase or decrease GHG emissions per unit of output. Rather, the principal effect of these provisions is on national output. If removing broad-based subsidies were offset by reducing distortionary taxes, the resulting increase in national output would be accompanied by increased GHG emissions. If the subsidies were replaced with lump-sum tax cuts that do not reduce distortions, there would likely be little effect on national output or emissions.

Sixth, it is difficult to estimate the impact of the broad-based tax expenditures on GHG emissions intensity. The committee examined the existing literature and commissioned modeling studies to estimate the effects of changes in the broad-based provisions on the overall GHG intensity of the economy. The results were not judged to be sufficiently reliable to draw firm conclusions.

Seventh, the effects of many tax provisions are complicated by their interaction with regulations. Very few tax provisions take place in a regulatory vacuum. Particularly in the energy sector, energy and environmental regulations overlay and interact with tax provisions. Prime examples are the interaction of highway motor fuels excise tax provisions with the CAFE standards for light-duty vehicles, the air pollution standards for the mix of electricity generation, the Renewable Portfolio Standards (RPS) for electricity generation, and the Renewable Fuel Standards for motor fuels blended from petroleum and ethanol. There are cases where regulations or mandates reinforce the effects of tax provisions and others where they offset their impacts. Analyses of the impacts of taxes on GHG emissions must take special care to include consideration of the regulatory environment.

Eighth, energy excise taxes reduce GHG emissions, but the impact is limited because of special features of the tax and because of regulatory constraints. The committee's estimates show unambiguously that highway fuel excise taxes reduce fuel consumption and GHG emissions. The analysis for this report finds that the current highway fuels taxes have a relatively small impact on GHG emissions because of the volumetric bias of the taxes as well as the constraints imposed by the renewable fuels standards.

RESEARCH RECOMMENDATIONS

The following recommendations to the Congress, the modeling community, the research support agencies, as well as the broader community provide guidance on the areas where the committee finds that more attention is needed. The committee recommends continued support of energy-economic modeling to better understand the impacts of taxes and other public policies on greenhouse gas emissions and the broader economy. Particular attention should be given to improving current models in the following ways:

First, models need to be made more transparent by clarifying both their assumptions and their structure; second, models should include measures of economic welfare that can be used to measure the efficiency and equity of policies; third, there should be more work to integrate partial equilibrium models with general equilibrium models so that the impact of revenue recycling and overall economic impacts can be more reliably estimated; and fourth, the committee recommends increased attention to studies that compare energy-economic models as a tool for improving understanding of models, narrowing the range of estimates, and improving model reliability.

GUIDANCE FOR SCORING GHG EMISSIONS

Because of the difficulties and resources required to provide reliable estimates, the committee discourages requiring the formal scoring of tax proposals for their impacts on GHG emissions. Much further work needs to be done before it can be accomplished routinely and reliably.

GUIDANCE FOR CLIMATE-RELATED TAX POLICY

In addition to estimating the impacts of the tax code on GHG emissions, the committee was asked to examine broader implications of taxes and climate-change policy. Although the committee does not make any recommendations about specific changes, the analysis undertaken for this report leads to several important insights and cautions about tax policy in the context of climate change.

First, current tax expenditures and subsidies are a poor tool for reducing greenhouse gases and achieving climate-change objectives. The committee has found that several existing provisions have perverse effects, while others yield little reduction in GHG emissions per dollar of revenue loss. The feedback effects within the energy sector (e.g., the fuel substitution effects when tax policy favors one source over others) or the international spillover effects (e.g., shifts in trade flows due to tax treatment differences) can offset or even reverse the expected direct effects of these policies. Such leakages and regulatory and tax arbitrage are common features of indirectly targeted provisions. Thus, if tax expenditures are to be made an effective tool for reducing GHG emissions, much more care will need to be applied to designing the provisions to avoid inefficiencies and perverse offsetting effects.

Second, some tax expenditures are more efficient than others. At their current scale, however, existing energy-related tax expenditures achieve small reductions in GHG emissions and are costly per unit of emissions reduction.

Third, the committee's reservations about tax expenditures and subsidies do not necessarily apply to tax incentives directly targeted on activities such as research and development on technological advances that will help the nation and the world transition to a low-carbon energy system.

Fourth, tax reforms that increase the economic efficiency of our economy may increase GHG emissions, but the increased output is likely much more than sufficient to pay for reducing the higher emissions as efficient climate-change policies are employed to reduce emissions.

Finally, a central finding of many studies in this area is that the most efficient way to reduce GHG emissions is through policies that create a market price for CO_2 and other GHGs. The committee finds that tax policy can make a substantial contribution to meeting the nation's climate-change objectives, but that the current approaches will not accomplish that. In order to meet ambitious climate-change objectives, a different approach that targets GHG emissions directly through taxes or tradable allowances will be both necessary and more efficient.

Chapter 1

Overview and Scope of the Study

INTRODUCTION AND ORIGIN OF THE STUDY

Legislative Background to the Study

In 2008, Congress directed the U.S. Department of the Treasury to work with the National Academies to undertake "a comprehensive review of the Internal Revenue Code of 1986 to identify the types of and specific tax provisions that have the largest effects on carbon and other greenhouse gas emissions and to estimate the magnitude of those effects."[1] Congress later appropriated funds so the study could commence in early 2011.[2]

In considering its task, the committee held discussions with staff of the Department of the Treasury, had an open meeting with the legislative sponsor of the study mandate, Representative Earl Blumenauer (D-OR), and held two public meetings to hear from interested parties. The committee also considered a suggestion from the staff of the Joint Committee on Taxation (JCT) that the study should "provide scientifically-based information to aid decision makers in the formulation of tax policies aimed at reducing emissions and mitigating climate change ... [and] ...identify the provisions of the Code that are most likely to have significant effects on carbon emissions" (JCT, 2009).

Those discussions led the committee to interpret Congress's request as including both provisions intended to affect energy-intensive activities in a narrow, specific way as well as provisions affecting major sectors of the economy, activities, or large segments of the population. We label the latter set "broad-based provisions" in this report. Moreover, we interpreted the charge to include any tax code provisions that might significantly affect emissions, whether by increasing or decreasing them, that is, whether they increase revenue (as in the

---

[1]Congress made its request as part of the Energy Improvement and Extension Act of 2008 (P.L. 110-343, Sec. 117).

[2]2010 Consolidated Appropriations Act (P.L. 110-343, Division B, Title I, Sec. 117).

case of excise taxes) or reduce revenues (as in the case of special deductions, exemptions, and credits that the Office of Management Budget [OMB] and the JCT call tax expenditures).

Limitations of the Study

The potential scope of the study is enormous because of the size and complexity of the Internal Revenue Code (IRC). Difficulties arise as well because it is difficult to gauge the full scope and depth of the mechanisms by which the tax code affects economic activity and greenhouse gas (GHG) emissions.

At the same time, there were clear boundaries on Congress's request. It did not ask the committee to assess the contribution of greenhouse gases to climate change, nor to examine the consequences of global warming, nor to recommend changes in specific provisions of the tax code.

Statement of Task

The National Academies' first task in carrying out its congressional mandate was to work with the Treasury Department to draft a statement of work reflecting an understanding of congressional expectations. The committee reviewed and accepted the following charge:

> The committee will undertake a consensus study to identify the types of and specific tax provisions that have substantial effects on the emission rates of carbon dioxide and other greenhouse gases, and to the extent possible rank the magnitudes of those effects.
>
> The committee will first determine the most appropriate analytical framework and methodology to use in examining the effects of the tax code on greenhouse gas emissions. It will consider both provisions that may increase emission rates as well as those having the effect of lowering them over specific periods, and both direct (e.g., fuel-related provisions) and indirect measures (e.g., the home mortgage deduction and the investment tax credit). Studying the tax code's impact on GHG emissions, the committee will necessarily focus heavily on energy, both the life cycles of different energy sources and their uses in different sectors such as electricity generation, transportation, industrial processes, and consumer uses (including in households). The study may extend to areas beyond energy, such as agriculture, forestry, urban development, and other land uses which can have significant effects on GHG emissions.
>
> The study will not recommend particular new taxes or tax incentives or changes in existing provisions of the tax code but may outline principles and criteria for formulating climate-sensitive tax policy in the future. It may evaluate the efficiency and effectiveness of different tax measures in reducing GHG emissions relative to other policy instruments.

STUDY CONTEXT: THE NATION'S ENERGY, FISCAL, AND CLIMATE-CHANGE DEBATES

We begin with a brief discussion of the context in which the study takes place. The issues addressed in this report lie at the intersection of three major ongoing public policy debates—those concerning national energy policy, fiscal and tax policy, and climate change policy. In each of these areas there are major controversies about the goals of policy, the relative priority of objectives, and the tools or policies that are best suited to attain the objectives.

Studying the tax code's effects on GHG emissions necessarily focuses on energy use, because a preponderance of emissions comes in the form of CO_2 released during the combustion of fossil fuels. The United States has struggled with energy policies for four decades, since the oil price shocks and embargo of 1973.

With respect to **energy policy**, some of the major issues are:

1. What are the external costs of energy—that is, what are the costs of energy use that are not reflected in market prices?
2. Should the nation take further steps to reduce the environmental and health damages from energy production and use, and, if so, which are the priority areas for policy?
3. What can be done to moderate the impact of higher energy prices and costs on consumers?
4. To what extent is security of supply a major concern, and in which energy markets?
5. To what extent should environmental policy rely on market-based instruments (such as pollution prices or taxes), and to what extent should the primary approach be regulatory standards?

Tax policy affects the allocation of goods and services, the distribution of income, and the servicing of the public debt. Among the key issues facing the nation here are:

6. How should measures to slow the rise in the federal debt be apportioned between tax increases and expenditure reductions?
7. Is there a role for environmental taxes in the mix of revenue increases?
8. Should the government eliminate or expand tax subsidies for fossil fuels and tax subsidies for conservation and renewable energy?
9. Should the government consider the effects on greenhouse gas emissions in designing broader tax subsidies, such as tax incentives for investment in machinery or tax subsidies for housing?
10. Is the use of earmarked taxes (such as the use of gasoline taxes to fund the highway trust fund) a sound approach to public finance?
11. To what extent should distribution of income impacts of the tax system be considered in its design or when proposing new tax provisions?

Finally, the debate about **climate-change policy** has emerged more recently, but has now joined debates about energy and tax policy on the national stage. Among the important questions here are:

12. What are future trends for climate change as well as the economic, human, and ecological impacts of those?
13. What are the most effective policy instruments to implement climate-change policies?
14. How much should climate-change policy rely on revenue-raising approaches, therefore contributing to deficit or tax reduction, compared to pursuing pure regulatory means?
15. Because climate change is a global problem, what approaches are best suited to implement policy on a multinational scale?

The committee summarized its task in the following way: To what extent do provisions of the U.S. federal tax code—with special attention to tax provisions focused on the energy sector—contribute in a negative or positive way to slowing the pace of climate change by affecting greenhouse gas emissions?

THE LINKAGES BETWEEN TAXES, ECONOMIC ACTIVITY, EMISSIONS, AND CLIMATE CHANGE

The issues discussed in this report involve the interaction of the tax code with the economy and the resulting impact upon the emissions of greenhouse gases and on climate change. We begin by discussing the U.S. federal tax system and particularly those parts that we have studied in this report. We explain how the tax system affects economic activity and thereby affects the emissions of GHGs and other forces that affect the global climate system. Finally, we provide a brief description of the scientific background on the role of emissions and other forces involved in changing climate.

Sources and Structure of U.S. Tax Law and Tax Expenditures

Sources of U.S. Tax Policy

The U.S. tax system arises from several sources of authority. Tax legislation embodied in the Internal Revenue Code (IRC) is the primary source. It outlines the individual income tax, which is the largest source of federal revenue, along with corporate income tax and excise taxes. Court cases also define important aspects of our tax system.[3] Treasury regulations interpreting the IRC and other administrative guidance constitute the balance of tax authority.

---

[3]For example, *Eisner v. Macomber*, 252 U.S. 189 (1920) and *Commissioner v. Glenshaw Glass*, 348 U.S. 426 (1955).

U.S. Tax Policy Structure, Function, and the Concept of Tax Expenditures

The main function of the tax system is to raise revenues to finance government operations and transfers. The choice of the level and structure of taxes also affects the stability of the economy, the distribution of household income, and medium- and long-term economic growth. Congress uses the tax system to promote social and economic goals through special tax reductions for some activities (called tax expenditures because they substitute for direct spending programs) and revenue-raising or excise taxes on other activities.

The key focus of this study is the last of these functions: how selected provisions of the tax code affect greenhouse gas emissions. While very few, if any, provisions in the tax code were explicitly, or as a matter of priority, designed to affect emissions, all provisions have at least an indirect impact. This study focuses on provisions meant to encourage or discourage activities that affect GHG emissions directly, especially energy-related activities, but also examines broad-based provisions that may have an indirect effect.

While the code primarily uses excise taxes as a means of discouraging selected activities (such as those imposed on the sale of highway motor fuels), tax rules can provide special tax benefits in a number of ways. For instance, the income from the activity may be excluded in whole or in part from taxable income, or tax may be imposed on the income at a preferential rate (i.e., a lower rate than for other income). Other provisions allow firms to deduct the capital costs of engaging in an activity in a single tax year instead of requiring the costs to be capitalized and recovered over the expected lifetime of the asset. Alternatively, the cost recovery deductions may be allowed on an accelerated schedule in comparison to the rate at which the asset declines in value. The code also provides tax credits for specific activities. Tax provisions granting special treatment to certain activities are substitutes for direct government expenditure of funds to support those activities. Hence, such provisions are often called tax expenditures.

The Congressional Budget Act of 1974 directed the newly established Congressional Budget Office (CBO) and the Office of Management and Budget (OMB) to publish annual lists of tax expenditures and estimates of their revenue cost.[4] The publication of an annual tax expenditure report for the U.S. Congress is now undertaken by the JCT. The Treasury creates an alternative set of calculations with a slightly different methodology, published annually as part of the President's budget documents prepared by OMB. Tax expenditures have continued to grow over the period since they were first tabulated. The JCT estimates

---

[4]Congressional Budget and Impoundment Control Act of 1974 [P. L. 93-344, Sec. 3(3)].

that tax expenditures grew from roughly $380 billion in 1974 to $1,100 billion for fiscal year 2011.[5]

Methodology of Calculation of Tax Expenditures

The Joint Committee on Taxation identifies tax expenditures as departures from a normative tax system and uses a specific method to compute expenditures.[6] Under the JCT staff methodology, the normal structure of the individual income tax includes the following major components: one personal exemption for each taxpayer and one for each dependent, the standard deduction, the existing tax rate schedule, and deductions for investment and employee business expenses. Most other tax benefits to individual taxpayers are classified as exceptions to normal income tax law.

Some scholars have criticized the JCT's methodology, because it compares the tax system actually in place against a hypothetical and subjectively determined tax system (a "normal" tax system) that has never existed. (JCT, 2008). We still used JCT's expenditure estimates in conjunction with estimates prepared by Treasury Department staff as an aid in deciding which IRC provisions to study for several reasons. JCT regularly computes and updates estimates and publishes details of its computational methodology. JCT's estimates are widely accepted, including by Congress, and rarely differ to any great extent from the other widely available set of estimates produced by the Treasury Department. Further, the consulting firms that carried out the modeling work regularly employ JCT estimates when analyzing policies.

Understanding the Link Between Taxation and GHG Emissions

Understanding the mechanism by which the tax code affects greenhouse gas emissions is central to the analysis and findings of this report. This mechanism operates primarily through the way taxes affect the prices of goods and inputs.

---

[5]The committee encourages readers to bear in mind that the Joint Committee on Taxation warns that because of the nature of expenditure estimates there is limited value in using total expenditure estimates for any year. All figures are in 2012 dollars. Summation and conversion to 2012 dollars performed by National Research Council staff.

Sources: The Staffs of the Treasury Department and Joint Committee on Internal Revenue Taxation. Estimates of Federal Tax Expenditures. JCS-11-75. July 8, 1975; Staff of the Joint Committee on Taxation. Estimates of Federal Tax Expenditures for Fiscal Years 2011-2015. JCS-1-12. January 17, 2012.

[6]A detailed explanation of JCT's methodology is contained in the annual tax expenditure report. See JCT, Estimates of Federal Tax Expenditures for Fiscal Years 2011-2015 (January 17, 2012). The definition of normal income is from the JCT's report for 2013.

The Three Routes from Taxes to Emissions

This study investigates the three major kinds of tax provisions. One set is excise tax provisions that apply to energy goods and services. Firms will pass on the cost of the excise taxes to consumers, raising the prices consumers pay. Thus, excise taxes on energy tend to increase the prices consumers pay, thereby decreasing consumption of energy. Less consumption generally means fewer GHG emissions, while more consumption would mean increased emissions.

A second set of tax expenditures affects inputs into the production process. For example, the oil depletion allowance is a tax expenditure that reduces the costs of producing petroleum and thereby encourages domestic oil production. If the increased production lowers world oil prices, consumption of oil and associated GHG emissions will increase. Similarly, tax provisions that encourage firms to increase the amount of electricity produced from renewables, such as wind power, should decrease GHG emissions, as those sources either displace fossil-fuel-generating capacity or avert construction of new fossil-fuel-generating capacity.

A final set of tax provisions, which we call "broad-based provisions" in this study, presents greater challenges. These include those that favor employer-provided health care, the deduction for home mortgage interest, and accelerated depreciation. The mortgage interest deduction, for example, lowers the cost of owner-occupied housing. This will increase housing consumption compared to consumption of other goods and services. If the increased emissions associated with housing are greater than the reduction in emissions from the goods displaced by the increased housing, then overall emissions would increase. Similarly, if increased housing consumption displaces more GHG-intense consumption, then these expenditures would lower overall emissions.

Figure 1-1 provides a schematic of how taxes affect GHG emissions. It shows that CO_2 and other GHGs are the outcome of a complex set of factors, of which tax provisions are just one.

General Equilibrium Considerations

One further issue must be considered in analyzing the effects of tax provisions. This is the "general equilibrium" impacts of tax changes. The idea is that we need to account for the fact that industries and households cannot be considered in isolation. When a tax provision affects one industry, it will also have effects on others.

For example, suppose that increasing the gasoline tax lowers the quantity of gasoline consumed, thereby changing GHG emissions from gasoline use. But the changes do not stop with the gasoline market. If the dollar value of gasoline

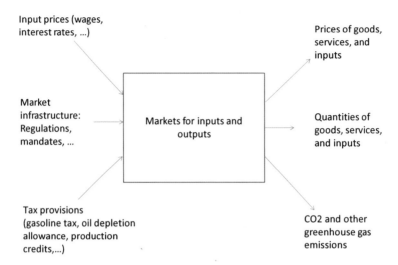

FIGURE 1-1 Schematic representation of how taxes affect GHG emissions through the market.

expenditures goes down, that will free up income for spending in other sectors. There will be secondary effects on output of goods and services in other sectors, depending on the impact on real wages, relative prices, and other factors. So we also need to consider the GHG emissions that arise from these ripple effects outside the gasoline sector. These ripple effects are called the "general equilibrium effects" of changing taxes.

Following this example further, perhaps people decide to fly rather than to drive (assuming the tax increase excludes aviation fuel). But air travel has relatively high GHG emissions as well, so there would be some offset to the reduction in the emissions from the gasoline sector by increases in GHG emissions in the airline sector. A complete analysis would incorporate all the sectors. The analysis should ideally also include effects in the rest of the world, since climate change is a global phenomenon, and because GHG emissions anywhere—not just in the United States—will affect future climate change. In other words, general equilibrium principles mean that a proper analysis must account for each sector in the entire economy, and in some cases the global impacts, in order to obtain a reliable estimate of GHG impacts.

Global focus of the impacts

A final reminder is that the focus of this study is ultimately on the global totals for consumption, production, and greenhouse gas emissions. While the charge was to examine the U.S. tax code, the impacts of the code are not limited to the U.S. borders. The U.S. tax system affects foreign as well as domestic production and consumption. Some provisions directly affect foreign economic activity through impacts on prices, exports, imports, or financial markets. Others have influences because the markets – such as petroleum or grains – are global in nature, so changes in domestic supply or demand spill over to other countries. Still others affect trade flows through tariffs and quotas. Some of the models used by the committee contain a full set of global linkages, and others are limited to U.S. production and consumption. While our analyses do not include every provision that affects global economic activity, the aim of the present report is to assess the impacts of the main provisions of the tax code that affect global greenhouse gas emissions.

The Link from Emissions to Climate Change

Although this report focuses primarily on the impact of the tax system on emissions of CO_2 and other greenhouse gases, Congress's ultimate concern in requesting this study is not with these gases per se, but with their impact on earth systems such as climate and also the health of oceans and biological systems. We conclude this chapter with a brief synopsis of the link from emissions to climate change and other earth systems to provide readers with a context for the discussion of the relationship between tax policy and GHG emissions. This discussion is based on the findings of many earlier peer-reviewed studies by the National Research Council, the United Nations Intergovernmental Panel on Climate Change, and a large body of research from scientific and academic institutions around the world. Each major study has concluded human-caused climate change is a real phenomenon well established by current knowledge about earth sciences.[7]

Scientists have established with high reliability that the basic aspects of global warming are taking place. We know from direct measurements of temperature that Earth's global mean surface air temperature has increased over the past century (IPCC, 2007). We know from direct measurements of atmospheric gases that greenhouse gas concentrations have also increased, and that this increase has been due primarily to emissions from human activities (IPCC, 2007). And we know that the observed global warming is consistent with the observed increase in greenhouse gas concentrations, but not with the background noise in the climate system, nor with changes in natural phenomena that can change Earth's climate, such as solar output, volcanoes, and Earth's orbital variations (IPCC, 2007; and Santer et al., 2013). While all measurements have some uncertainty, each of the above conclusions is an established scientific fact.

---

[7]See National Research Council, 1979, 1983, 1992, 2002, 2010, and 2011 for some of the most important NRC reports. Other important surveys are the Intergovernmental Panel on Climate Change, 1990, 1996, 2002, and 2007.

Greenhouse Gases and Earth's Energy Balance

Greenhouse gases are necessary for life as we know it on Earth. Indeed, if Earth had no atmosphere to absorb a portion of the energy radiated out due to the heating of the Earth's surface, the surface temperature would be approximately -18°C (Kasting and Catling, 2003), well below the freezing point of water. However, Earth's surface receives energy input not only in the form of short-wave radiation from the Sun but also in the form of long-wave radiation that is radiated from the atmosphere back down to Earth's surface. Calculating the energy balance for Earth with this atmospheric "greenhouse effect" shows that greenhouse gases are responsible for elevating and maintaining Earth's surface temperature from well below the freezing point of water to the mild and habitable global mean surface temperature of approximately 15°C that is observed at present (Kasting and Catling, 2003).

The potency of these greenhouse gases arises from the fact that they are essentially transparent to short-wave radiation but absorb a substantial fraction of long-wave radiation. As a result, short-wave radiation emitted from the Sun passes through these gases, while long-wave radiation emitted from Earth's surface is partially absorbed, providing additional energy input to Earth's surface. Thus, although greenhouse gases are present only in trace amounts in the atmosphere, they have a substantial influence on Earth's surface temperature.

Observed Changes in Greenhouse Gas Concentrations

Greenhouse gases include water vapor, carbon dioxide, methane, nitrous oxide, and fluorinated gases such as the hydrofluorocarbons that are used as refrigerants in appliances such as refrigerators or air conditioners. While some greenhouse gases have natural sources, many greenhouse gases are also emitted by human activities, including carbon dioxide emissions from combustion of fossil fuels, nitrous oxide emissions from agricultural fertilizer use, and methane emissions from leaks in natural gas pipelines. Analysis of the atmospheric chemistry of gases indicates that the atmospheric concentration of many greenhouse gases has been increasing since the start of the industrial revolution (around 1750), and that human activities are primarily responsible for that increase (IPCC, 2007).

The U.S. Environmental Protection Agency tracks emissions from different sources in the United States and reports them annually in its Greenhouse Gases Inventory (EPA 2010). These are combined using a formula for "carbon dioxide equivalence," which estimates each compound's effect on the Earth's energy balance over a 100-year period. In 2010, U.S. greenhouse gas emissions were 84 percent carbon dioxide, 10 percent methane, 4 percent nitrous oxide, and 2 percent fluorinated gases using this CO_2-equivalent. For 2010, 34 percent were from electricity production, of which 70 percent is supplied by fossil fuels; 27 percent from transportation, of which 90 percent is supplied by petroleum-based fuels; 21 percent from industry, including both energy consumption and industrial chemical reactions; 11 percent from commercial and residential activities such as building heating and cooling; and 7 percent from agriculture, including methane from livestock and nitrous oxide from fertilizer application and other practices. In addition, land use and forestry, which can act as a source or sink of CO_2, has been a net sink in the United States, offsetting about 15 percent of U.S. emissions in 2010. Given these contributions, the committee must consider a suite of different gases, not just CO_2, and different economic activities in order to meet its charge.

Global Warming: Observed Changes in Earth's Surface Temperature and Potential Impacts of Continued Global Warming

Scientists have reconstructed measures of the global mean surface temperature based on instrumental records. Global mean surface temperatures have increased over the past century by about 0.8°C, most rapidly over the past four decades (IPCC, 2007). Independent analysis by several research groups from multiple countries reveals very similar calculations of global-scale temperature change over the past century (IPCC, 2007).

Earth's climate has changed dramatically over geological history (IPCC, 2007), and an important question involves the role of human factors in recent warming. Climate scientists estimate the impacts of greenhouse gases on past and future climates using computerized numerical climate models. These have been developed over more than a half-century using well-established physical laws, such as the laws of conservation of momentum, mass, and energy, as well as based on detailed representations of the Earth's geography. Climate models have been validated by comparing model calculations with the observed climate record. Thus, scientists can use the models to make hypotheses about the strength of background noise in the climate system, as well as how climate responds to changes in factors such as increased concentrations of greenhouse gases, volcanic eruptions, and the strength of solar output. Comparisons of model calculations and historical data indicate that the observed magnitude and pattern of warming in the atmosphere and ocean are not consistent with the background noise in the climate system, nor with observed changes in solar output or volcanic eruptions, but are consistent with measured anthropogenic increases in atmospheric greenhouse gas concentrations (IPCC, 2007; and Santer et al., 2013).

Although the cause of recent and rapid climate change cannot be directly observed, the many published studies of detection and attribution show that human activities are the predominant reason for observed global warming, and for a number of related observed climate changes. Model calculations consistently show that further increases in global greenhouse gas concentrations are virtually certain to cause further global warming. Model experiments also show that regional climates are highly likely to change in response to this global warming.

Most model runs indicate that a future with higher greenhouse gas concentrations will be one with more energy in the atmosphere, likely leading to changes in regional climates, increasing occurrence of extreme hot events, increasing intensity of precipitation, increasing lengths of dry spells, and decreasing snowpack (IPCC, 2012 and IPCC, 2007).

THE CLIMATE CONTEXT OF THE COMMITTEE'S CHARGE

The robust scientific understanding of the connection between greenhouse gas emissions, global warming, and regional climate change provides the broad context for the committee's charge. The contribution of increasing atmospheric concentration of greenhouse gases to Earth's changing climate is well understood, as is the connection between a changing global climate and many of the changes in regional climate, including changes in the frequency and magnitude of extreme weather events. The well-established climate changes that have already occurred—and those that are likely to occur in the future should greenhouse gas concentrations continue to rise—motivated the congressional direction to undertake this study, and thereby provide the context in which the committee has executed its charge.

Chapter 2

Methods for Evaluating Tax Policy Effects on Greenhouse Gas Emissions

INTRODUCTION

The last chapter described the background for the present study. This chapter examines the connection between tax policy and greenhouse gas (GHG) emissions along with the approach the committee followed. We begin by explaining the mechanism by which tax policy can affect greenhouse gas emissions and then explain how the committee chose specific provisions of the Internal Revenue Code (IRC) for in-depth analysis. We also give an overview of the concept of tax expenditures, because our methodology for choosing provisions depended on estimates of tax expenditures. Another important facet of the analysis is the regulatory environment and how regulations outside of the tax code affect the code's impacts on emissions.

We next provide a brief description of economic modeling and how the absence of available literature on this topic led the committee to commission new economic modeling to analyze the greenhouse gas impacts of select tax code provisions. In the last part of the chapter, we provide a lengthier discussion of how we chose the four models we used, give some details of how those models represent the economy, and then discuss some of the parameters of the models. A more complete discussion is contained in Appendix A.

HOW TAX CODE PROVISIONS IMPACT GREENHOUSE GAS EMISSIONS

Taxes are one of the many factors that affect the level and composition of economic activity, which in turn determines the level of anthropogenic emissions of CO_2 and other greenhouse gases. The purpose of this study is to untangle the effects of different individual tax provisions from other determinants of output and to estimate how through these changes the tax code affects U.S. and global GHG emissions.

The percentage depletion allowance provides an illustrative example of the task before the committee. The tax code provides two ways for firms to recover capital expenditures made to establish a mineral property such as an oil or gas well. Under cost depletion, each year the owner of the well can deduct from taxable income a portion of the costs of developing the well. Over the life of the well the total amount deducted cannot exceed the total amount invested. Percentage depletion allows owners a deduction based on a percentage of that year's gross receipts from the well, and the total lifetime amount can exceed the cost of developing the well. Currently, the code allows producers of oil and gas to deduct 15 percent of the gross income of a property when calculating taxable income. The special provision is subject to several limitations (such as being available only to independent producers, only up to 1,000 barrels per day, and limited to 50 percent of net income). According to the Joint Committee on Taxation (JCT), this deduction reduced tax revenues by as much as $0.9 billion in fiscal year 2011.[1] The question asked of the committee is, how much did this change the emissions of CO_2 and other GHGs?

At first glance, the impact of this policy on GHG emissions appears straightforward. By allowing a larger deduction for depletion than the actual economic costs, the percentage depletion allowance subsidizes oil and gas production. It thereby lowers prices for petroleum products and increases petroleum use and the resulting GHG emissions. A first estimate could look like the following. One could attempt to estimate using econometric methods how much production of oil increased because of the percentage depletion allowance. Multiplying that amount by the emissions per unit of oil production would then yield a first-order estimate of the amount of additional emissions resulting from the tax preference.

In reality, estimating the impact would be much more difficult than appears at first glance. In practice, the information on covered oil production would be difficult to obtain. The Internal Revenue Service does not release the necessary data from corporate income tax returns. Nonintegrated firms may have production and output records, but such records are unlikely to reveal how much additional oil the firms produce on account of the incentive. Second, the provision applies to natural gas as well as oil, and it would be necessary to include the impacts on gas in the calculations. Third, the substitution between oil and gas and other fuels would need to be considered. Fourth, since oil is bought and sold in world markets, the role of global supply and demand would need to be considered.

Further complicating the issue, climate change results from GHG emissions anywhere around the globe and thus depends upon total world consumption of fuels. From the standpoint of a U.S. inventory of greenhouse gas emissions, we may be interested in how a tax provision affects U.S. fuel

---

[1] The JCT estimates revenues foregone of $0.9 billion, while the Treasury Department estimates $1.2 billion.

consumption. But for climate change, we need to consider how any provision would ripple outside the United States and affect consumption abroad as well.

The complexities do not end there. If depletion allowances subsidize oil and natural gas production, but the produced gas displaces coal use, then even though the subsidy is contributing to the production of a fuel that will release GHGs when burned, it may actually be lowering total emissions from what they otherwise might have been, because coal is more GHG intensive than natural gas. This point is particularly important for provisions that favor non-carbon-emitting sources like wind. They contribute to lower emissions only to the extent they on balance displace GHG-emitting technologies. So the GHG reduction benefits depend on what they displace or induce.

Calculations of emissions impacts must also incorporate the upstream and downstream impacts of any changes. Biofuels provide a useful example. It is tempting to suppose that biofuels would have zero or very small net GHG emissions, because they come from plant material that will be regrown at the next harvest. Recent research has shown that this supposition is far from the mark (see the discussion in Chapter 5). Production, transport, and conversion of biofuels all require energy input, usually from fossil fuels. Fertilizers used for crop production require energy to produce, and nitrogen fertilizer directly produces nitrous oxide emissions, itself a powerful greenhouse gas. Further, conversion of non-crop land to biofuels crops may reduce the existing stock of carbon.

The effects of broader provisions in the tax code introduce yet further complications. Accelerated depreciation or provisions affecting housing directly affect decisions on investment and housing, but just how those are related to greenhouse gas emissions is far from obvious. Such provisions may affect the overall level of national output and through that channel affect emissions. Tax code provisions also affect the mix of goods and services produced among industries and thus would alter emissions per unit of total production if some sectors are more emissions-intensive than others. Additionally, some of the provisions have effects on the pattern of housing choices and on land use and real estate development. It is tempting to ignore these economy-wide ripple effects as unimportant. But tax changes, if applied to a very large economic base, may have large impacts on GHG emissions. We discuss such potential impacts of broad-based tax preferences in Chapter 6.

APPROACH AND METHODOLOGY OF THIS STUDY

Given limited time and resources, the committee needed to decide which of the many provisions in the tax code to analyze. We reasoned that the largest effects were likely to result from some combination of provisions with large tax revenue effects and those targeted toward activities that were closely related to greenhouse gas emissions. Of course, without actually making the calculations of the impacts, the committee could not know how to combine these two factors to develop a firm ranking. Further, the complexity of interactions of regulations, markets, and technology often means this assumed relationship does not necessarily hold. However, we faced the need to limit the size of task to available time and resources.

This study could not proceed in the way that many National Academies' studies do, with the committee's task largely confined to evaluating and synthesizing an existing body of literature to answer the questions presented in the statement of task. While there is an extensive literature on the response of economic activity to various tax incentives, including those related to energy provisions, with only a few exceptions do existing studies carry through to the impact on greenhouse gas emissions. Thus, the committee faced a task of estimating the GHG impacts of the tax code provisions virtually from scratch.

Because of the complexities discussed above, the committee chose to use economic and energy models to evaluate the impacts of different provisions. These many complexities have a critical implication for the strategy that the committee pursued in answering its charge. It is clear that the only way to assess the implications of the tax code for greenhouse gas emissions is to put together a comprehensive or integrated framework that assesses each of these effects and how they interact, moving from the specific tax provision to how that provision affects markets and then to how these market changes affect the many decisions across the economy that produce greenhouse gas emissions. The comprehensive framework used in this study is integrated energy-economic and greenhouse gas modeling. Models are numerical representations of a sector or an economy that attempt to capture the most important economic relationships and behavioral responses that generate these effects under consideration. The models the committee chose to use for this study incorporate estimates of the response of firms and individuals to tax incentives and incorporate the impacts of these behavioral changes on GHG emissions.

The committee then directed the National Academies to contract with research teams who have developed and used these integrated energy-economic models to calculate the effects of changes in different tax provisions on GHG emissions. This chapter's section on modeling approaches and comparative strengths of models considered for use in supporting analyses and Appendix A review the models considered, those for which it was possible to develop timely analysis, and a comparison of the relative strengths and weaknesses of different models.

There are literally dozens of integrated energy-economic models that are currently in use around the world. These differ in terms of their structures, algorithms, databases, empirical estimations, sectoral details, and economic comprehensiveness. Appendix A provides more details about the models used in this study, including their structure and capabilities, documentation, and alternatives that the committee considered. In addition, because the tax provisions span such a wide range of economic sectors—ranging from very narrow ones directly affecting only a small component of the energy sector to those that affect nearly every decision in the economy—no single model could investigate all of the provisions we targeted for analysis. Nevertheless, for most of the provisions we wanted to investigate for this study, we identified at least one model capable of analyzing it. Moreover, there was some overlap among models so that we could compare results across models.

MAJOR TAX PROVISIONS SELECTED FOR ANALYSIS

Given the size and complexity of the U.S. tax code, it is clearly impossible to analyze more than a small fraction of its provisions. The committee's statement of task delimits the task as follows: The committee "will undertake a study to identify tax provisions that have substantial effects on the emission rates of carbon dioxide and other greenhouse gases."

Given the charge, the committee interpreted its task as evaluating specific provisions of that code that are closely related to incentives for activities that would likely affect emissions or ones that have been discussed as a target for change and could have large indirect effects on emissions. After reviewing the existing literature and the major provisions, the committee selected tax provisions based on three criteria:

1. The provisions would include the IRC sections most likely to have a significant impact on greenhouse gas emissions.
2. The provisions could be analyzed using available economic models.
3. The analysis would provide guidance on how to undertake a similar analysis of proposed tax legislation in the future for those provisions we could not investigate for this report.

This section reviews the method and the rationale the committee used to decide which provisions to analyze as part of this study.

Excise Taxes

Excises are taxes on the sale of specific goods or when certain activities are undertaken. For example, the federal government and most state governments charge a per-gallon excise tax on gasoline for highway use. The federal government also imposes taxes on air transport of people or property. As shown in Table 2-1, 61 percent of all federal excise revenues came from taxes on transportation fuels or transportation activities. Given the large portion of excise revenue from these fuel taxes, the committee decided to consider the top three of these—the taxes on highway gasoline and diesel fuels, the airline passenger tax, and the taxes on aviation fuels. Ultimately, modeling capabilities allowed analysis only of the tax on motor fuels for highway use and the tax on aviation fuel, shown in bold in Table 2-1. Moreover, we found a large literature discussing the motor fuels excise taxes' impacts on fuel consumption and vehicle miles traveled along with one study that attempts to assess the tax's impact on greenhouse gas emissions. We discuss the literature and the findings from our commissioned modeling analysis of energy-sector excise taxes in Chapter 4.

TABLE 2-1 The 10 Largest Excise Tax Collections for Fiscal Year 2010

Excise Tax	Amount Collected FY 2010 (billions of $)[2]	Chapter Where Discussed
Gasoline	**25.1**	**Chapter 4**
Tobacco, domestic	15.9	N/A
Diesel fuel, except for trains and intercity buses	**8.6**	**Chapter 4**
Transportation of persons by air	7.6	N/A
Liquor, domestic	3.7	N/A
Beer, domestic	3.2	N/A
Use of international air travel facilities	2.4	N/A
Truck, trailer, and semitrailer chassis and bodies, and tractors	1.9	N/A
Liquor, imported	1.3	N/A
Telephone and teletypewriter services	1.1	N/A
Aviation fuel	**0.4**	**Chapter 4**

Bold entries were analyzed as part of this study.
Source: Internal Revenue Service, Statistics of Income Bulletin Historical Table 20: Federal Excise Taxes Reported to or Collected by the Internal Revenue Service, Alcohol and Tobacco Tax and Trade Bureau, and Customs Service, Fiscal Years 1999-2010.

Tax Expenditures

Every year the staffs of the congressional Joint Committee on Taxation (JCT) and the Treasury Department (Treasury) prepare an annual compendium of tax expenditures, the latter for the administration's budget by the Office of Management and Budget.[3] The Congressional Budget Act of 1974 defines tax expenditures as "revenue losses attributable to provisions of the Federal tax laws which allow a special exclusion, exemption, or deduction from gross income or which provide a special credit, a preferential rate of tax, or a deferral of tax liability."

---

[2]Internal Revenue Service Statistics of Income Division. *Federal Excise Taxes Reported to or Collected by the Internal Revenue Service, Alcohol and Tobacco Tax and Trade Bureau, and Customs Service, by Type of Excise Tax* (Table 20). All amounts are nominal dollars.

[3]JCT's estimates are reported in JCT ESTIMATES, usually published in January each year. The Treasury estimates are published in Office of Management and Budget's annual budget analysis (*Analytical Perspectives*). The two use different methodology to prepare the estimates. For a discussion of the differences, see JCT ESTIMATES, pp. 23-24.

Traditionally, the government has seen this foregone revenue as equivalent to subsidizing the activity directly through a budget expenditure. JCT and Treasury organize each set of estimates by budget function activity, and report estimates for the current fiscal year and the next 4 years. These analyses include provisions that are scheduled to expire, permanent provisions, and some provisions that have expired but nonetheless continue to have revenue impacts.

From the long lists of tax expenditures, the committee sought to analyze those likely to have the largest impacts on greenhouse gas emissions. As mentioned in Chapter 1, combustion of fossil fuels accounts for roughly 90 percent of anthropogenic greenhouse gas emissions[4] (U.S. Environmental Protection Agency, 2012). The committee therefore looked first at IRC provisions that directly affect energy production or consumption. For energy-related tax expenditures, the committee ranked each provision by the size of the estimated revenue change for the year 2010.[5]

This ranking revealed that the 10 policies that the Treasury estimates result in the largest amounts of estimated foregone revenue combined account for over 90 percent of all estimated tax expenditures due to energy-specific code sections. Because of this, the committee initially considered those 10 as a first step to winnow the field. Table 2-2 lists these 10 largest energy-specific tax expenditures. Bolded entries indicate provisions that the committee was able to analyze with models as part of this study. Chapters 3 and 5 analyze the impact of these energy-sector tax policies on GHG emissions.

Broad-based Tax Expenditures

In addition to provisions that directly affect the energy sector, the committee examined broad-based provisions that may indirectly affect emissions. Table 2-3 shows the 10 largest broad-based tax expenditures for fiscal year 2010. After examination and discussion with modeling teams, the committee chose three provisions for the study: subsidies to housing, subsidies to health care insurance, and accelerated depreciation of machinery and equipment.

---

[4]Percent share of energy-related emissions computed by committee and National Research Council staff.

[5]The 2010 figures were the most current that were available at the time the committee began its deliberations. The ordering and composition of the 10 largest tax expenditures estimates did not change appreciably through the study duration with the notable exception of the Volumetric Ethanol Excise Tax Credit, which expired in 2012. Many of these energy-related tax subsidies were created to support fossil fuel industries, because they were seen as key to war efforts during World Wars I and II. Fossil fuel industries continued to receive the bulk of energy tax expenditures until the turn of the twenty-first century when provisions favoring renewable fuels began to grow. As of this writing, the tax subsidies now favor renewable energy sources over fossil fuels in terms of total cost to the Treasury.

TABLE 2-2 The 10 Largest Energy Tax Policies (by dollars of foregone revenue)

Provision	FY 2010 (billions of 2010$)[6]	Chapter Where Discussed
Alcohol Fuel Credit and Excise Tax Exemption	**5.75**	**Chapter 5**
Credit for Electricity Production from Renewable Sources (Including the cash grant in lieu of tax credit)	**3.90**	**Chapter 3**
Credit for Energy Efficiency Improvements to Existing Homes	3.19	Chapter 3
Excess of Percentage over Cost Depletion for Oil and Gas Wells	**0.98**	**Chapter 3**
Special Tax Rate on Nuclear Decommissioning Reserve Funds	0.90	Chapter 3
Temporary 50-Percent Expensing for Equipment Used in the Refining of Liquid Fuels	0.76	N/A
Biodiesel Producer Tax Credit	**0.51**	**Chapter 5**
Expensing of Exploration and Development Costs for Oil and Gas	0.40	N/A
Credit for Investment Renewable Energy Infrastructure	0.30	N/A
Tax Credit and Deduction for Clean-burning Vehicles	**0.24**	**Chapter 3**
Preferential Tax Treatment of Certain Publicly Traded Partnerships with Qualified Income Derived from Certain Energy-related Activities	0.50 (JCT)	N/A
Credits for Advanced Energy-manufacturing Facilities	0.18	N/A

Some policies are codified in multiple IRC provisions. **Bold** entries were analyzed using models as part of this study.
Sources: Office of Management and Budget, Fiscal Year 2012 Analytical Perspectives, Budget of the U.S. Government. Staff of the Joint Committee on Taxation, Estimates of Federal Tax Expenditures for Fiscal Years 2010-2014.

---

[6]The figure we report here for Alcohol Fuel Credits includes the refundable excise tax credit mentioned in Treasury's footnote 2. Likewise, the figure for the electricity production from renewable resources includes the cash grants offered by the Treasury in lieu of the production credit mentioned in Treasury's footnote 1. All amounts are nominal dollars.

TABLE 2-3 The 10 Largest Broad-based Tax Expenditures

Provision	FY 2010 (billions of 2010$)	Chapter Where Discussed
Exclusion of Employer Contributions for Medical Insurance Premiums and Medical Care	**160.1**	**Chapter 6**
Deductibility of Mortgage Interest on Owner-occupied Homes	**79.1**	**Chapter 6**
Earned Income Tax Credit	59.6	N/A
Exclusion of 401(k) Plans	52.2	N/A
Accelerated Depreciation of Machinery and Equipment (normal tax method)	**39.8**	**Chapter 6**
Exclusion of Employer-sponsored Retirement Plans	39.6	N/A
Step-up Basis of Capital Gains at Death	39.5	N/A
Making Work Pay Tax Credit	38.9	N/A
Deferral of Income from Controlled Foreign Corporations (normal tax method)	38.1	N/A
Capital Gains (except agriculture, timber, iron ore, and coal)	36.3	N/A

Source: Office of Management and Budget, Fiscal Year 2012 Analytical Perspectives, Budget of the U.S. Government.

While the broad-based tax provisions do not affect the energy sector or GHG emissions directly, they may have an effect on overall emissions, because they affect output in large portions of the economy. There are two routes by which broad-based tax provisions can affect GHG emissions. First, they may change the mix of goods and services produced from high (or low) GHG-intensive sectors to low (or high) GHG-intensive sectors, thereby affecting over-all emissions. Second, they may affect the overall size of the economy and therefore change emissions simply because the economy is larger or smaller.

At $160 billion per year, the exemption for employer-provided health in-surance premiums is the largest federal tax subsidy. Even though this provision is not directed toward energy per se, because health spending is such a large part of the overall economy, a change in its size or energy use could have a signifi-cant impact on GHG emissions.

Similarly, the housing subsidies are available to a large number of taxpay-ers. Moreover, the links between housing and emissions are more straightfor-ward than the links for health care. Because the subsidy lowers the cost of hous-ing, it makes it easier for families and individuals to own more or larger houses.

Residential housing directly accounts for one-fifth of all U.S. GHG emissions and is involved indirectly through development patterns that may reduce housing density, thereby increasing emissions from automobiles (U.S. Environmental Protection Agency, 2012). A change in the amount of housing may therefore have sizeable impacts on GHG emissions.

Accelerated depreciation allows firms to deduct their investments in machinery and equipment on a faster schedule than the standard tax lifetime. This tax preference lowers the cost of capital and encourages firms to invest more, thereby resulting in some combination of increased amounts and faster turnover of the affected capital. There are multiple ways in which energy use and greenhouse gas emissions would be affected by this provision. A larger capital stock will increase output, and, other things equal, increase emissions. Newer capital may be more energy efficient and a bigger share of it will decrease average energy intensity, or reduce energy use and emissions to the extent it replaces older inefficient capital. Finally, a lower cost of capital would favor the growth of capital-intensive sectors of production, which could increase or reduce energy use and emissions depending on whether capital and energy are complements or substitutes in production. The effects of changing broad-based provisions are analyzed in Chapter 6.

REGULATORY INTERACTIONS

Although other laws and regulations are not specifically included in the committee's charge from Congress and the Treasury, they are a critical element because they interact with tax laws to influence GHG emissions. In some cases, regulations can reduce the overall impact of tax provisions on GHGs, while in other cases they may increase the impacts. For example, the federal government has instituted a Renewable Fuel Standard (RFS) mandate that requires transportation motor fuel sold in the United States to contain a minimum volume of renewable fuel. Separately, until 2012 sections 40 and 40A of the IRC offered tax credits to producers of renewable fuels as an incentive to increase production and use of these fuels. Since the RFS mandate sets a minimum annual production quantity of fuel, producers would produce at least that much fuel each year even if the tax incentive did not exist. Thus, the tax incentive must operate within the constraints set by the minimum production mandate.

Another example of regulatory interaction is the Corporate Average Fuel Economy (CAFE) standards. These require new automobiles to attain a given average miles per gallon for each manufacturer. As these standards become increasingly tight, the impact of gasoline taxes will generally be reduced because fuel consumption starts from a lower base. However, as fuel economy standards are tightened, studies need to incorporate the indirect impact or rebound effect, which in this context is a phenomenon whereby the increased fuel efficiency induces car owners to drive more and therefore use more fuel.

In addition to federal regulations such as the RFS mandate and CAFE standards, the committee needed to consider the effects of direct subsidies as well as state and local regulations, subsidies, and tax policies. To the extent practicable, all existing laws and policies of different governments were incorporated into the baseline assumptions used in the computational models in order to include their effects when estimating the impacts of federal tax provisions.

AVAILABLE LITERATURE ON EFFECTS OF TAX PROVISIONS ON GREENHOUSE GAS EMISSIONS

An important first step in analyzing the potential impacts of the tax code on emissions and climate change was a review of the existing literature. The committee and several commissioned consultants undertook a systematic review of studies to analyze the impact of taxes and subsidies on GHG emissions. Although there is a vast literature on tax expenditures, there is virtually no empirical research on the impacts of the U.S. tax code on GHG emissions. We found some studies that perform one or more of the steps in the analysis outlined above, but only one that provides empirical estimates of the impacts for most of the energy-sector tax expenditures considered in this report.

This literature search confirmed that there are many studies of the demand and supply for gasoline; of the effects of subsidies on the diffusion of renewables; of the effects on GHG emissions of tax subsidies for biofuels; of the impact of accelerated depreciation on economic growth; and of the impact of housing and health care deductions on tax revenues and the composition of economic activity. There is also a vast array of econometric studies of demand and supply elasticities for fuels and electricity. The literature is large even if we focus on some of the major survey articles (e.g, Bohi, 1981, 1984; Dahl C., 1993, 2002, 2012; Dahl and Duggan, 1988; Ko, 2001; Espey, 2004; Taylor, 1975, 1977; Wade, 2003).

Little to no literature exists, at any level of analysis, for other tax provisions of interest. We found no published empirical studies of the impact on energy use of the nuclear decommissioning tax preference. The committee found a single unpublished working paper (Hitaj, 2012) that provides an econometric analysis of how production tax credits for renewable electricity affect wind capacity in the United States. However, to estimate the impact of the credits on total GHG emissions, the estimates in this paper would need to be supplemented by estimates of how changes in wind generation affect electricity markets and other energy markets as a whole.[7]

---

[7]Palmer et al. (2010) in a background paper for the Resources for the Future-National Energy Policy Institute (RFF-NEPI) study use the National Energy Modeling System (NEMS) model to analyze the CO_2 implications of the production tax credits (PTC) and the investment tax credits (ITC). The RFF-NEPI study analyzes the impacts of regula-

Similar problems exist in the case of credits for energy-efficient improvements to homes. There is a considerable literature on the impact of price incentives on the purchase of energy-efficient appliances and home improvements, but none that examines the impact on the investments targeted by these credits.[8] It is also necessary to estimate the impact of energy-efficient improvements on energy demand, the consequent effects on energy markets, and the corresponding changes in GHG emissions. Again, we found only the one previously mentioned study that does this.

The only area where there is a substantial literature is on the impact of gasoline taxes on fuel consumption and GHG emissions. This literature includes studies using one of the computational models employed in this study (Krupnick, 2010)[9] as well as econometric studies of vehicle ownership and gasoline demand (see Bento, 2009, and Gillingham, 2011). Many of these studies do not, however, incorporate important features of the U.S. tax and regulatory systems, such as biofuels taxes and subsidies, CAFE standards, and regulatory mandates for ethanol and other biofuels. Nor do they include the general equilibrium effects. Instead, these studies focus on estimates of supply and demand elasticities and how a change in the tax-inclusive price of fuels and electricity will affect demand and supply. While any estimate of the impact of changes in a tax incentive must begin with these, that is just a first step toward estimating GHG-emission impacts.

Researchers at the University of Nevada, Las Vegas's Center for Business and Economic Research (CBER) authored the one study the committee reviewed that contained a comprehensive analysis of the effect of government energy subsidies. This study analyzed the effects of many tax expenditures on CO_2 emissions, but did not include other GHGs (Allaire and Brown, 2011). CBER modeled the effects of prices and taxes on energy markets using a simplified supply-and-demand framework. We found this study to be a useful first-order analysis, and compared its results to those from the work we commissioned using models with more detailed representations of energy markets, technologies, and regulations to ensure that we captured a full picture of the impacts of tax changes.

ECONOMIC EFFICIENCY AND ECONOMY-WIDE MEASURES TO REDUCE GHG EMISSIONS

Another set of studies puts the economic and GHG impacts of tax policy in the broader context of climate-change objectives. There is a substantial literature that investigates approaches to achieving national and global climate-

---

tions to promote energy efficiency and renewable energy use on CO_2 emissions, but does not examine the effects of tax expenditures.

[8]The credits are authorized by sections 25C and 25D of the tax code.

[9]The U.S. Energy Information Administration's (EIA) National Energy Modeling System, described in full detail in Chapter 3.

change objectives (such as concentrations or temperature limits) in the most efficient way.[10] The strategy in these studies is to determine the least-cost approaches to achieving a given objective. Generally, these studies use either marketable GHG emissions permits or taxes on GHG emissions as the actual mechanisms to achieve the targets cost-effectively.

Both analytical studies and empirical studies find that the most efficient approach to reducing emissions is through uniform or economy-wide coverage of taxes or regulations.[11] The efficient trajectories are ones in which the marginal costs of emissions reductions are equalized in every sector of the economy (and indeed in different countries for global efficiency). This condition would imply, under certain standard economic assumptions, that the price of emissions (in the case of tradable allowances) or tax per unit of CO_2-equivalent emissions (in the case of GHG emissions taxes) is uniform in every sector. A uniform carbon price provides appropriate incentives for consumers, producers, entrepreneurs, and innovators to adjust their activities so as to reduce emissions and encourage low-emissions technologies in the most efficient manner.[12]

One of the important features of uniform carbon pricing is that the policy directly targets GHG emissions rather than indirectly targeting capital goods, processes, or products that are only indirectly linked to emissions. Studies have also found that the cost per unit of emissions reductions is higher—often much higher—when sector-specific tax expenditures, subsidies, or regulations are used than when economy-wide measures are employed (EIA, 2011; NRC, 2010; Clarke, et al, 2009; Krupnick, 2010).

Put differently, a key finding of economic studies of climate-change policy is that the most reliable and efficient way to achieve given climate-change objectives is to use direct tax or regulatory policies that create a market price for CO_2 and other greenhouse gas emissions.

The economic advantage of targeted policies was emphasized in the National Research Council report *America's Climate Choices,* in its overall summary:

> Emission reductions can be achieved in part through expanding current local, state, and regional-level efforts, but analyses suggest that the best way to amplify and accelerate such efforts, and to minimize overall costs (for any given national emissions reduction target), is with a comprehensive,

---

[10]See William J. Baumol; Wallace E. Oates, *The Theory of Environmental Policy*, 2nd Edition, Cambridge University Press, 1988; Robert Stavins, "Transactions Costs and Tradable Permits," *Journal of Environmental Economics and Management*, 29, 1995, 133-148.

[11]See for example, Intergovernmental Panel on Climate Change, Third Assessment Report, *Mitigation*, Chapter 6, and specifically p. 413, for a discussion of different instruments.

[12]A useful survey of the literature with an analysis of different approaches is in International Energy Agency, *Energy efficiency policy and carbon pricing,* Information Paper, Energy Efficiency Series, International Energy Agency, Paris, 2011.

nationally uniform, increasing price on CO_2 emissions, with a price trajectory sufficient to drive major investments in energy efficiency and low-carbon technologies. (p. 3)

This was further elaborated:[13]

Most economists and policy analysts have concluded, however, that putting a price on CO_2 emissions (that is, implementing a "carbon price") that rises over time is the least costly path to significantly reduce emissions and the most efficient means to provide continuous incentives for innovation and for the long-term investments necessary to develop and deploy new low-carbon technologies and infrastructure. A carbon price designed to minimize costs could be imposed either as a comprehensive carbon tax with no loopholes or as a comprehensive cap-and-trade system that covers all major emissions sources. (p. 58)

The use of uniform carbon pricing has been the organizing principle of the European Union's Emissions Trading Scheme (ETS), which has operated effectively for almost a decade.

A recent study by Resources for the Future (RFF) was particularly useful for the committee's analysis, because it provided quantitative estimates of the relative efficiency of different policy instruments (Krupnick, 2010). Moreover, the study used one of the models employed by the committee (NEMS model) and the same consultant to undertake the modeling (OnLocation, Inc.).

The RFF study examined the resource costs of reducing CO_2 emissions over the 2012-2030 period using different regulatory approaches. (In nontechnical terms, resource costs refer to the losses in real income of the country.) It estimated that the average resource cost is $12 per ton of CO_2 reduced (2007 USD) for a carbon tax that is set at $18 per ton of CO_2 in 2012 and rises to $67 per ton of CO_2 in 2020. (All figures in this paragraph are in 2007 USD.) Such a tax is estimated to reduce CO_2 emissions 40 percent from 2005 levels by 2030. The economic cost of a cap-and-trade that auctions the allowances is the same (ignoring complications about uncertainty of policy). In contrast, the cost per ton of CO_2 reduced of all the other policies examined was higher. For example, the resource cost per ton of CO_2 reduced by implementing the building codes for new residential construction described in the Waxman-Markey bill of 2009 was estimated to be $51 per ton. The resource cost of reducing emissions by the PTC and ITC was estimated at $34 per ton of CO_2. Therefore, each of the subsidies

---

[13]The report particularly cited C. Fischer and R. G. Newell, "Environmental and technology policies for climate mitigation" (*Journal of Environmental Economics and Management* 55(2):142-162, 2008); and T. H. Tietenberg, *Emissions Trading: Principles and Practice* (Washington, D.C.: Resources for the Future, 2006).

and sector-specific policies were more costly than a uniform national policy of raising carbon prices.[14]

The same finding has been emphasized in several reports and literature reviews. These have consistently found that the approach of uniform pricing of GHGs is the most reliable and efficient way to achieve different climate-change objectives (EIA, 2011; NRC, 2010; Clarke, et al, 2009).

MODELING APPROACHES AND COMPARATIVE STRENGTHS OF MODELS CONSIDERED FOR USE IN SUPPORTING ANALYSES

The committee's review of the literature determined that there is a very thin body of published work on which to base an analysis of the impacts of the tax code on greenhouse gas emissions. In reality, there are virtually no studies that analyze the impact in an empirical analysis that reflects the full complexity of the U.S. energy system and regulatory environment. Given the lack of existing work, the committee decided to direct the National Academies to undertake studies with existing best-practice empirical models. The next section discusses the committee's approach in detail.

To undertake the new modeling efforts for this study, the committee decided the most sensible approach was to work with existing modeling groups rather than attempting to create one or more new models. Many models already exist that link energy markets with environmental components. We selected models with long track records of use for academic research, for public policy analysis, and by private enterprises, each of which has been carefully and repeatedly scrutinized by the research community through a lengthy history of peer-reviewed publications. Additionally, we wanted to avoid any conflicts of interest and therefore did not choose models that committee members were actively managing. We also had to work within the constraints of the budget and time line of the National Academies' contract with the Department of the Treasury and thus needed to identify models and contractors that could deliver results on time and within budget.

As mentioned in this chapter's introduction, no one model or type of model can adequately analyze all of the provisions we wished to investigate. Our review of existing models suggested we would need to utilize one each of three types of models suitable to our task:

---

[14]Note that these estimates refer to the analysis "with no market failure." The study also examined the cost-effectiveness of policies "with market failures." These included particularly the possibility that households might apply an inappropriately high discount rate to energy-efficiency investments. The study found that one specific technology (geothermal heat pumps) had favorable costs relative to other technologies, but the committee did not investigate this proposal.

1. Models focused on energy markets and representing them in considerable detail;
2. Models focused on agricultural markets that included detailed representation of how biofuels policies would affect those markets; and
3. Economy-wide models that often include specific but less detailed representation of energy markets, agricultural markets, or both, along with details of a few tax policies.

Appendix A offers further detail on the options within each class of model and their principal features. In the following section, we explain our choice of models used and their chief characteristics.

Energy-sector Models

We considered three energy-focused models.[15] Ultimately we chose to use two of these. For the bulk of our analysis, we employed the U.S. Energy Information Administration's (EIA) National Energy Modeling System (NEMS), run by a professional consulting firm.[16] We also used an energy-sector model developed at the Center for Business and Economic Research at the University of Nevada, Las Vegas (CBER).

Details of the NEMS Model

The committee looked to NEMS for the bulk of our analysis here for several reasons. The U.S. Energy Information Administration of the U.S. Department of Energy designed, implements, and continues to maintain NEMS. EIA publicly publishes details of the model structure and assumptions and updates the model annually to incorporate energy market data from the prior year.[17] EIA also uses NEMS to produce its *Annual Energy Outlook*, an analysis and projection of energy market trends, typically over a 25-year period. Because of these efforts by EIA, NEMS's capabilities and shortcomings are well understood within the energy-modeling and energy-economics communities.

NEMS's representation of energy markets focuses on four interactions: (1) energy supply-energy conversion-energy demand, (2) domestic energy system-economy, (3) domestic energy market-world energy market, and (4) economic

---

[15]These were the MARKAL model, which is more modeling framework that can be developed for specific applications depending on interests and data availability; the NEMS model, developed and used by the U.S. Energy Information Administration in its annual energy outlook and available for other uses; and the CBER model, developed at the University of Nevada, Las Vegas.

[16]OnLocation, Inc., of Vienna, Virginia.

[17]EIA provides documentation on the NEMS model on its Web site: http://www.eia.gov/analysis/model-documentation.cfm.

decision making over time. There are many important assumptions that drive the NEMS model, the two most important being U.S. economic growth and oil prices. Other assumptions include macroeconomic and financial factors, world energy markets, resource availability and costs, behavioral and technological choice criteria, cost and performance characteristics of energy technologies, and demographics.

NEMS consists of four supply modules (one for each major fuel), two conversion modules, four end-use demand modules, one module to simulate energy-economy interactions, one module to simulate international energy markets, and one module that provides the mechanism to achieve a general equilibrium among all the other modules. These details make NEMS an ideal, widely available model for analyzing the energy-focused provisions we sought to study. Importantly, though, CO_2 is the only GHG that NEMS includes. Given the model's focus on energy markets, this is understandable, but means it could not represent potentially important effects of biofuels policies on methane or land use.

Details of the CBER Model

We also asked UNLV's Center for Business and Economic Research to extend a prior analysis it performed using its simplified supply-and-demand model of the energy sector (Allaire and Brown, 2011). In CBER, demand and supply are represented as functions of energy prices (including the prices of substitutes and complements).[18] The key parameters in the CBER model are the price elasticities of supply and demand. The elasticities for the CBER model are derived by expert judgment of the authors from reviews of the economic literature.

An important advantage of the CBER model is that it was developed with an objective of investigating the effects of subsidies, mostly tax expenditures, on CO_2 emissions in the energy sector. The developers invested considerable effort in representing most of the energy-specific tax code provisions. Indeed, their earlier paper is the only comprehensive study of the impact of the tax code on CO_2 emissions.

An additional strength of the CBER study is that consumer behavioral responses and broad technological trends are represented to the extent these can be estimated from historical data. However, the model has important shortcomings compared to all the other models used in this study. First, it is a "static" model, meaning that everything takes place as an equilibrium in a single stylized time period. This means that it cannot capture the dynamics of capital turnover. It

---

[18]The researchers involved solved the equations in their model using a standard mathematical solver package. The equations can be found in the following report: M. Allaire and S. Brown (August 2012). U.S. Energy Subsidies: Effects on Energy Markets and Carbon Dioxide Emissions. Retrieved 2013, from http://www.pewtrusts.org/uploaded Files/wwwpewtrustsorg/Reports/Fiscal_and_Budget_Policy/EnergySubsidiesFINAL.pdf.

also provides no detailed representation of the regulatory environment necessary for examining some provisions, such as those involving the Renewable Fuels Standard for biofuels, and so does not capture many important effects of those regulations on energy markets. Lastly, CBER's model includes only one GHG, CO_2. While CO_2 is the largest GHG by total emissions, this means the model can not capture potentially important changes in methane or land use resulting from biofuels policies.

Details of the FAPRI Model

The committee's focus on the agricultural sector stemmed largely from tax provisions related to biofuels. There are several models that have been used to examine the economics of biofuels and biofuel policy.[19] Analysis of these policies is extremely complex.

Challenging aspects of modeling biofuel policy include (1) the complex interactions with agriculture and agricultural policy, including competing demands for crops and by-products supplies of animal feeds; (2) the complex policy requirements of the Renewable Fuel Standard (RFS2, as described below); (3) investment and production tax credits that differentially treat different biofuel production pathways and feedstocks; (4) international linkages in agriculture and energy markets; (5) land-use change and competition for land between agriculture and other uses of land; and (6) the implications of land-use change for GHG emissions.

Given consideration of budget and ability of different groups to produce timely analysis, we decided that Missouri University's implementation of the Food and Agriculture Policy Research Institute's model, FAPRI-MU, best fit the requirements.[20] This model has the combination of detailed agriculture and crop

---

[19]These include macroeconomic models such as Emissions Predictions and Policy Analysis (EPPA) and Global Trade Analysis Project (GTAP) models (e.g., Gurgel, Reilly, and Paltsev, 2007; Gurgel et al., 2011; Tyner et al., 2010; Decreux and Valin, 2007). Agricultural optimization models including the Forest and Agricultural Sector Optimization Model (FASOM) (Adams et al., 1996, as in Beach and McCarl, 2010; Beach, et al., 2013); simulation models such as MiniCAM (Wise et al., 2009); and econometric-based simulation models such as the Food and Agricultural Policy Research Institute (FAPRI) model (Babcock and Carriquiry, 2010).

[20]FAPRI provides documentation on its model on the FAPRI-MU website: http://www. fapri.missouri.edu/outreach/publications/umc.asp?current_page=outreach. See FAPRI-MU Report #12-11, Model Documentation for Biomass, Cellulosic Biofuels, Renewable and Conventional Electricity, Natural Gas and Coal Markets; FAPRI-MU Report #09-11, FAPRI-MU Stochastic U.S. Crop Model Documentation; FAPRI-MU Report #05-11, New Challenges in Agricultural Modeling: Relating Energy and Farm Commodity Prices; FAPRI-MU Report #09-10, FAPRI-MU U.S. Biofuels, Corn Processing, Distillers Grains, Fats, Switchgrass, and Corn Stover Model Documentation; FAPRI-MU Report #07-08,

markets specification, linkage to international markets, inclusion of regulatory constraints relevant for analysis of biofuel provisions, and tax parameters that allow an analysis of changes in the various parameters. It also has a full representation of the intricacies of renewable fuel credits, with multiple fuel-production pathways representing both conventional and second-generation processes tied with links to global markets for crude petroleum and refined fuels.

However, FAPRI-MU does not explicitly consider land use or the carbon implications of land-use change, a potentially large and important emissions pathway. Land-use changes are highly uncertain, with wide-ranging results found in the literature (Plevin et al., 2010; Searchinger et al., 2008; Melillo et al., 2009; Keeney and Hertel, 2009; Hertel, 2011; Tyner et al., 2010; and Mosnier et al., 2012). Instead, greenhouse gas implications are assessed by applying a fixed GHG coefficient per unit of fuel for different biofuel production pathways. Some coefficients include a factor to estimate land-use change implications on emissions. Default estimates are those of the U.S. Environmental Protection Agency (2010) that include CO_2, N_2O, and CH_4 implications of land-use change.

Details of the IGEM Model

In principle, economy-wide general equilibrium models are potentially capable of analyzing many of the tax provisions, but they are limited in that they lack the granularity needed for some of the detailed provisions. For example, an economy-wide model that represents the electricity sector as single production function cannot accurately represent the effect of a provision directed at individual technologies such as wind, solar, or nuclear. And, similarly a model that simplifies the agricultural sector as producing a single product is less able than a detailed model of the agricultural sector, with a full range of crops and livestock, to trace how a biofuels policy may affect corn production and land-use change. So the economy-wide models are most useful for considering the broad-based tax provisions and for capturing general equilibrium effects of tax policies, but are less useful for capturing sectoral details.

The committee considered six economy-wide models that could analyze broad tax provisions.[21] Given the aims of the study and the constraints, the

---

Model of the U.S. Ethanol Market; and FPARI-UMC Report #12-04, Documentation of the FAPRI Modeling System.

[21]These included the MIT Emissions Predictions and Policy Analysis (EPPA) model (Paltsev et al., 2009) and/or the MIT U.S. Regional Energy Policy (USREP) model (Rausch et al., 2010) very similar to EPPA but with greater detail on the United States. The Applied Dynamic Analysis of the Global Economy (ADAGE) model (Ross, et al., 2009) developed at RTI International and widely used by the Environmental Protection Agency for analysis of greenhouse gas policies (e.g., EPA, 2009). The Multi-Region National (MRN) model developed at Charles River Associates (Berstein, et al., 2007). The Global Trade Analysis Project (GTAP) model developed at Purdue University

committee decided that the Intertemporal General Equilibrium Model (IGEM)[22] was best positioned to address the questions involved. The committee chose IGEM because it is a well-established economy-wide model containing detailed specification of the U.S. tax code, because several agencies of the U.S. government have used it for many years for energy and environmental modeling, and because the firm that runs it was able to deliver modeling results in a timely fashion.

IGEM is a multisector general equilibrium model that represents the economy following modern neoclassical economic theory. It captures relationships between industry outputs and final consumption goods as represented in input-output tables for the U.S. economy and includes an expanded social accounting system with estimates of factor returns from each production sector and the disposition of goods to final-demand sectors (households, government, investment, and exports). Thus, an important feature of IGEM is that it captures the effects of tax provisions in one industry on the output and emissions in other related sectors. For example, a provision in the steel industry will have effects on the automobile industry, and that effect can in principle be captured in a general equilibrium model. IGEM also can be used to illustrate how a tax change that alters incentives to work, save, and invest would affect the overall level of economic activity.

The IGEM model differs from most other models, as it includes a time series of data for the U.S. economy, and most parameters are econometrically estimated from these data.

Models differ in how they deal with the dynamics of economic adjustment, that is, how investment decisions react to changing prices and interest rates and to expectations of future conditions. IGEM is a deterministic forward-looking model, as are all other models used in the present study. In other words, it assumes that individual consumers and firms look forward and anticipate future conditions with perfect foresight, making decisions today based on those expectations. Agents are said to have perfect foresight because their expectations are realized exactly. This means that the model assumes that firms and consumers know the trajectory of oil prices, GDP growth, and other important factors. Forward-looking models clearly overestimate the capability of agents to look forward, but the implications of this assumption for estimating the impacts of tax policy on GHG emissions is unclear.

IGEM has a representation of the capital stock in which capital is fully malleable or adaptable to changing circumstances. This is equivalent to assuming that firms can quickly and economically retrofit their capital—such as buildings, machinery, or equipment—in response to changing market and regulatory conditions.

(Hertel, et al., 2010). And the Intertemporal General Equilibrium Model (IGEM) of the United States (Goettle et al., 2007), developed at Dale Jorgenson Associates.

[22]Dale Jorgenson Associates provides documentation on the IGEM model on its website: www.igem.insightworks.com/.

IGEM has a very detailed disaggregation of production sectors, but within the constraints of the Standard Industrial Classification system. In this classification, for example, electricity generation is a sector, but there is no further distinction between renewable power sources (solar, wind, nuclear, hydro) and nonrenewable sources of generation (coal, gas, or oil). Fuels, capital, labor, and intermediate goods are inputs into the electricity sector, and some amount of electricity is produced. Implicitly, substitution of capital for fuels could be interpreted as an increase in one of the non-fossil-fuel technologies. Rates of technical change are econometrically estimated based on the historical data, and these are represented as time trends on input requirements rather than explicit technologies.

IGEM has a more extensive representation of the U.S. tax system than most other economy-wide models. This feature was important for consideration of the broad provisions. IGEM does not contain as much detail of energy or agriculture markets as the models we chose specifically to model biofuels and energy-focused provisions, and hence it was not able to simulate many of those narrow provisions. On the other hand, the model does include a comprehensive suite of GHGs including CO_2, CH_4, N_2O, and so-called high global warming potential gases, or HGWP. We discuss further details of the IGEM simulations used in this study in Chapter 6.

GOVERNMENTAL POLICIES TO PROMOTE INNOVATION AND LOW-CARBON TECHNOLOGIES

The economic analysis of subsidies and tax expenditures for standard goods and services like shoes and pizzas finds that they lead to economic distortions and reduce national output and economic welfare. When there are no market failures, subsidies may distort prices and outputs away from their market-determined levels. This is a standard finding about competitive markets (Debreu, 1959).

However, investments in innovation and new technologies suffer from a market failure because of the inability of inventors to appropriate the full value of their activities (Arrow, 1962). Policies such as rights to intellectual property and government support for research and development can improve innovative performance and lead to increases in national output. Particularly in cases where markets do not reflect true social costs, as is the case with emissions of GHGs, research support can play a vital role in promoting low-carbon technologies.

These analyses lead to the important conclusion that subsidies and other policies to support new technologies are a critical component of a strategy to slow climate change and do not have the same inefficiencies that are found in subsidies of standard goods and services. These points were emphasized in the recent report on U.S. climate policies by the National Academies. This report concluded as follows with respect to governmental support of energy technologies:

Major technological changes in the U.S. energy system and other sectors will be needed to reduce GHG emissions significantly, and this will require an infusion of financial and human resources to support each phase of the process.... Resources that are critical for technology innovation include money for R&D and people with the requisite training, skills, and creativity to innovate. (NRC, 2010)

INDUCED TECHNOLOGICAL INNOVATION,
THE TAX SYSTEM, AND ECONOMIC MODELS

A final issue arising in evaluating the impact of taxes on GHG emissions is their impacts on technological change and innovation. This is particularly important for the long run. For example, a tax or subsidy that favors renewable technology X would lower the relative price of X and raise the production and use of X. In anticipation of a larger market, firms would devote more resources to research, development, and commercialization of X. Additionally, there might be learning by doing, through which higher levels of cumulative investment and production lower the production costs of the technology. These processes are often called "induced innovation."

Evaluating the rate and direction of technological change has proven a major challenge. It involves predicting the outcomes of research and development that have not yet been completed, or perhaps even funded. Methods necessarily extrapolate from historical experience and assume that past results are an indicator of future performance. Studies on induced innovation have developed two leading theoretical frameworks. The "research framework" of induced innovation arose in an attempt to understand why technological change appears to have been largely labor saving (Hicks, 1932; Nelson, 1959; Arrow, 1962). More recently, this approach was further developed as the new growth theory (Romer, 1990; Aghion and Howitt, 1999). Under the research approach, higher levels of investment in knowledge will expand society's production possibilities and increase the long-run growth rate of the economy. Over several years, Jorgenson and co-authors have adopted the research approach and currently incorporate it as part of their econometrically estimated production structure in IGEM (see, for example, Jorgenson and Wilcoxen, 1991).

The alternative approach to modeling-induced innovation is the "learning model." This approach has become particularly widely used by energy-economic models in recent years as models increase the granularity of the technological description down to individual technologies. The learning approach is included in the NEMS model. For example, most electricity technologies are assumed to have a cost decrease of 1 percent for every doubling of cumulative capacity in the late stages of development, while the learning rate ranges from 5 to 20 percent in the early stages (EIA, 2010). A more detailed discussion of learning in the NEMS model is provided in Chapter 3.

After reviewing the existing literature, the committee concludes that the impact of induced innovation is clear: Tax provisions that lower the cost or raise the production of a specific technology will generally tend to lead to induced innovation and improve that specific technology. However, while the sign of the impacts is clear, the size of the impact and the impact on GHG emissions is highly speculative. Moreover, the mechanism by which these would work will differ greatly depending upon which of the different innovation models is assumed to be driving the impact from technological change.

While clearly an important topic, estimating the impacts of the tax code on innovation is a problem beyond the capability of current empirical models. The influence of learning is highly controversial, and there are no reliable structural models of learning in the energy sector. There is a substantial literature on the impact of changing energy prices and other prices on patents (see, for example, Popp, 2002, 2004; Popp and Newell, 2012; Popp, Hascic, and Medhi, 2011). There is also work on the impact of energy prices on productivity (see, for example, Newell et al., 1999). However, existing empirical studies of induced innovation in the energy sector do not extend to the effects of changes in the tax code. Additionally, they do not estimate the general equilibrium impact of induced innovation on outputs in different sectors or to the impact on GHG emissions. A few studies examine the impact of policies such as carbon taxes or cap and trade on economy-wide technological change and innovation (see Popp, 2004; Bosetti et al., 2006). However, these studies are generally at a highly aggregated level and do not focus on specific provisions of the tax code.

Given these difficulties, the committee did not undertake a separate study of the effects of the tax code on emissions through induced innovation. Nor did the committee attempt to separate out the influence of endogenous technological change or learning from the other forces at work, the most important being substitution. Part of the reason for the decision not to pursue this area is that estimating induced innovation empirically has proved very challenging. Additionally, this would have required another set of model runs, along with major model modifications, and each would have required time and budgets well beyond what we had available. Finally, there are no widely accepted models available that adequately represent the daunting complexities that arise with endogenous changes in technology. Because of the importance of technological change for emissions reductions, particularly over the long run, we point to work on induced innovation as an area of particular interest for further research and improvement of energy-economic models.

TAX BASELINE FOR MODELING AND ANALYTICAL APPROACH

Baseline Assumptions

In addition to choosing which models to use and which provisions to analyze, the committee needed to select a baseline to use as a comparison for evaluating the effects of changing tax policies on greenhouse gas emissions. A baseline is a set of economic, tax, and regulatory assumptions that is used as a starting point for analyses. Given a baseline, a model can change certain assumptions and determine the impact on important outcomes, such as output or emissions. In the present study, the major changes were those involving provisions of the tax code.

The committee determined that the most suitable baseline for its task was the tax code and regulatory system of 2011, with all provisions extended indefinitely. Additionally, the committee adopted for its economic assumptions, such as GDP growth rate and global petroleum prices, the assumptions used by the EIA in its *Annual Energy Outlook 2012*. The major advantage of this approach is that it represents an actual tax and regulatory system, and that could be used as a starting point for comparative and counterfactual calculations. Under this baseline, each model would first estimate energy use and emissions under the baseline assumptions. For comparative purposes, provisions would then be removed from the tax code and all model outputs, including emissions, would be re-estimated with the difference attributed to the provision under consideration. Note that not all models were able to incorporate exactly the same baseline assumptions, and the differences are noted below in the discussion of the individual models and provisions.

Treatment of Revenue Changes

Another important issue that required attention was to determine how to treat changes in government revenues that would follow changes in tax provisions. Removing provisions from the tax code will change the revenues coming to the government. Eliminating a tax expenditure would raise revenue, while removing an excise tax would reduce revenue. Modeling the full effects of a change in the tax provisions requires assumptions about what the government does with increased revenues and how it pays for reduced revenues. Among the several alternative approaches discussed below, the committee adopted three in the calculations.

Assume Revenue Changes Have No Effects

The simplest approach is to assume that the revenue changes have no effect on the economy. This approach avoids making any assumption about how the government would adjust its budget in response to revenue changes, and examines the economic and emissions effects of a policy change absent any fiscal changes.

This approach is implicitly embodied in the partial equilibrium models. These models simply take the revenues and send them into the rest of the economy, but no impacts on output, income, or prices in the rest of the economy are included. Excluding fiscal impacts can lead to misleading results, particularly for policy comparisons. For example, a large increase in a very efficient tax can look more costly than a tiny increase in a very inefficient tax, but it might be a more efficient policy because it raises more revenue. We note the disadvantage of not treating revenues correctly, but for some provisions there is no alternative to partial equilibrium models.

Allow the Budget Deficit to Change in General Equilibrium Calculations

The simplest approach is unsatisfactory given that changes in government revenues must affect either borrowing or spending. Economy-wide models that include the dynamics of taxes, expenditures, and debt can in principle consider the treatment of changing revenues, since they consider all markets in the economy, estimating emissions in a general equilibrium, where demand and supply are in balance in every market within the economy. Ignoring revenue changes would mean that the government budget no longer adds up in simple accounting terms. Therefore, in a dynamic general equilibrium model, the model must make some assumption about how the rest of the government budget adjusts to accommodate the revenue change.

Offset Revenue Changes Using Lump-sum Payments

The committee directed the IGEM team to recycle revenues in two different ways: through lump-sum changes in the tax system and through proportional changes in tax rates.

A lump-sum change is one that changes taxes or government transfer payments by a fixed amount that does not depend on household behavior. For example, it could take place through a fixed-dollar refundable tax credit available to all households.

This approach is computationally simple. However, in reality, the government generally does not change taxes in a lump-sum manner. Instead, it collects revenue via distortionary taxes such as the income tax or taxes on companies. Economists refer to "distortionary" taxes as those that alter relative prices of goods, capital, or labor, thereby causing people to change their economic decisions from what would otherwise have been optimal choices in order to reduce their tax liability. Income taxes distort the choice between labor and leisure (or market work and home production) by reducing the relative return from an extra hour of work. They distort the choice between present and future consumption by altering the rates of return on different assets. Selective excise taxes distort consumer choices by raising the relative prices of taxed goods.

When household and firms change their behavior to reduce their tax liability (for example, by working less, saving less, or buying fewer taxed goods), the result is a reduction in economic well-being. Therefore, raising government revenue from taxes that distort behavior imposes costs on the private sector that exceed the amount of revenue collected; this is labeled an "excess burden" of taxation. Assuming that all revenue losses are offset by lump-sum taxes implicitly ignores the losses (or excess burden) from distortionary taxes, while assuming that all revenue gains are returned to taxpayers with lump-sum subsidies ignores the extra benefit that could arise from reducing distortionary taxes.

Offset Revenue Changes by Adjusting Tax Rates

An alternative approach is to assume that revenue changes induced by changing a tax provision are offset by raising or lowering marginal tax rates. The committee asked the IGEM contractor to implement this approach by raising or lowering all corporate and individual tax rates by the same proportional amount. For example, if the top marginal tax rate of corporations is 35 percent and individual rates range from 10 to 35 percent, a 10 percent proportional cut would reduce the corporate rate to 31.5 percent and individual rates to a range of 9 to 31.5 percent. This approach is used because it is transparent and easily modeled.

Comparing Approaches: Offsetting Revenue Changes via Lump-Sum Payments v. Adjusting Tax Rates

While the two approaches to offsetting revenue changes (lump-sum changes and proportional changes in tax rates) were not the only possibilities, they are relatively easy to implement and interpret. The main reason for choosing the lump-sum approach is it allows analysts to focus on the effects of the energy tax change by itself as nearly as possible. The main advantage of the tax-rate-change assumption is that it allows us to capture the effects on distortions of scaling up or down the main taxes used to fund general government services in our current tax structure in a manner similar to how revenues are currently raised.

Table 2-4 details what provisions were able to be studied in this report, what chapter discusses the analysis for specific provisions, and what models, if any, were used for analysis.

MEASURES TO APPRAISE MODEL ESTIMATES

The committee was particularly attentive to determining the reliability of the estimates from the modeling teams. Each of the three major modeling teams that were engaged by the committee has a long track record of work in energy modeling for the U.S. government and for scholarly journals. Notwithstanding past work, the committee took steps at successive stages of the study to determine the appropriateness of the modeling approach, the assumptions, and the model outputs as well as to understand the intuition behind the model results. These steps reflected the need to follow a rigorous standard of quality control for a topic of such importance for public policy and for understanding the impacts on government revenues and climate change.

TABLE 2-4 Provisions Modeled for This Study and Where Discussed in This Report

Provision	Chapter Where Discussed	Model(s) Employed in Analysis
Credit for electricity production from renewable sources (including the cash grant in lieu of tax credit)	Chapter 3	NEMS-NAS
Excess of percentage over cost depletion for oil and gas wells	Chapter 3	NEMS-NAS
Credit for energy efficiency improvements to existing homes	Chapter 3	Models not Used for Analysis
Special tax rate on nuclear decommissioning reserve funds	Chapter 3	Models not Used for Analysis
Excise tax on highway motor fuels (gasoline and diesel)	Chapter 4	NEMS-NAS; CBER; FAPRI-MU; IGEM
Excise tax on aviation fuel	Chapter 4	CBER
Alcohol fuel credit and excise tax exemption	Chapter 5	NEMS-NAS; FAPRI-MU
Biodiesel producer tax credit	Chapter 5	NEMS-NAS; FAPRI-MU
Exclusion of employer contributions for medical insurance premiums and medical care	Chapter 6	IGEM
Deductibility of mortgage interest on owner-occupied homes	Chapter 6	IGEM
Accelerated depreciation of machinery and equipment (normal tax method)	Chapter 6	IGEM

Independent Verification of Model Choices, Baseline Assumptions, and Tax Code Provisions Analyzed

As a first step, before entering into contracts with the modeling groups selected, the committee sought an external analysis of our choices of models, modeling assumptions, and criteria for tax provisions to analyze. We asked two experts from the economics and tax policy communities to undertake this task. They were Dr. William Pizer, Associate Professor of Economics at Duke University, recognized expert on economics and energy modeling, and former deputy assistant secretary for environment and energy in the U.S. Department of the Treasury; and Dr. Richard Newell, Associate Professor of Energy and Environmental Economics at Duke University and former director of the U.S. Energy Information Administration, the agency that is responsible for official U.S. government energy statistics and analysis.

In March 2012 Pizer and Newell each submitted independent written analyses of our methods and preliminary choices. These reports endorsed the committee's decisions on models and specifications. At the same time, they made several important suggestions for improvements and modifications. On April 10, 2012, they discussed their reports in a joint conference call with the committee.

Both consultants agreed that the models that the committee had selected were the most appropriate among available options. Among their specific suggestions for modeling strategy were suggestions for the analysis of biofuels subsidies. Following their recommendations, the committee added the credit and deduction for clean-burning vehicles and the credit for advanced energy-manufacturing facilities. How to handle the revenue changes that accompany a change in tax policy was of particular concern to both reviewers. Our choice to request that IGEM analyses be performed both with the revenue being offset by changing tax rates and by lump-sum rebates arose in part from our discussions. These exchanges are part of the record of the committee's deliberations, and their implications are explained in the following chapters.

A second part of the committee's quality assurance involved working closely with the modeling teams to ensure that the baseline assumptions were correctly and consistently incorporated in their models. Individual members of the committee with experience in the types of models being used worked closely with the modeling groups between committee meetings. The committee member biographies in Appendix B describe this expertise in each case.

Finally, as the results were reported to us, the full committee carefully scrutinized the simulation results for anomalies and possible errors. In those instances, we asked the contractors to

- Elaborate their procedures for estimating errors;
- Explain why their model produced an unexpected result;
- Provide a procedure to decompose emissions into sources (e.g., for IGEM);
- Confirm the input parameters;
- Rerun a simulation if questions remained; and
- Conduct sensitivity analyses of key model parameters and input assumptions.

For NEMS, we relied on the contractor, OnLocation, Inc., which has reprogrammed the NEMS model and therefore represents an independent validation. Because the CBER model was extremely simple, we tested some of the results using auxiliary calculations. The committee spent most of three of our five committee meetings and several conference calls between meetings analyzing the modeling results, deliberating about their reliability, validity, and interpretation, and articulating their limitations.

The committee recognizes that the quality control procedures that it followed were just a subset of possible steps in error estimation and validation of complex models. We could not check the tens of thousands of lines of code, data points, and calculations of the different models; nor could we retest and re-estimate the econometric studies. To do so would have extended the time and scope of the study indefinitely and in any case greatly exceeded our financial resources. Rather, in the following chapters, the committee describes the issues that arise in interpreting the model results as well as the confidence that should be placed in the results.

SUMMARY

The committee commissioned four modeling groups to conduct new studies investigating the implications of tax code provisions on GHG emissions. The models were diverse in their construction, data, time periods, complexity, regional coverage, tax structure, and degree of coverage of the overall economy. The models employed to the maximum feasible extent a common baseline assumption for tax, regulatory, and economic scenarios. While none of the models could calculate a full range of estimates for all provisions, most major provisions were analyzed, and some were analyzed by multiple models.

The models have different strengths and weaknesses. The committee concluded that the NEMS model was best suited for most of the narrow provisions directed toward the energy sector. The FAPRI model had distinct advantages for analysis of biofuels tax incentives among those we studied. The IGEM model was the only model able to consider most of the broad-based tax provisions analyzed by the committee and to weigh economy-wide impacts. The CBER model was useful for providing a comprehensive analysis of the GHG impacts of provisions as well as for providing comparisons with other models.

Chapter 3

Energy-Related Tax Expenditures

PLAN OF THE CHAPTERS ON SUBSTANTIVE PROVISIONS

Earlier chapters have outlined the scope of the present report as well as the approach that the committee has taken in addressing its charge. Chapters 3 through 6 present the detailed analysis of the impacts of different tax provisions on greenhouse gas (GHG) emissions. As was described in the last two chapters, the bulk of the results are based on specific modeling for this report undertaken by four external contractors.

The plan is the following. The present chapter examines the impact of the major energy-related tax expenditures. Chapter 4 reviews energy-related excise taxes. Chapter 5 analyzes the subsidies and regulations affecting biofuels, a subset of the energy-related tax expenditures characterized by significant subsidies and a particularly complex set of markets and regulations. Finally, Chapter 6 examines a number of broad-based tax expenditures to determine whether they may have a significant impact on greenhouse gas emissions.

Each of the chapters has a parallel structure. It describes the major tax expenditures, describes the modeling efforts undertaken by the modeling groups, explains the results, and then presents the overall conclusions. Chapter 7 then presents an overall summary of the results along with findings and recommendations.

ENERGY-RELATED TAX EXPENDITURES

In 2011 the 10 largest tax expenditures that directly affected the energy sector resulted in a loss of $16.9 billion in tax revenues (see Chapter 2, Table 2-1). The largest of these, the alcohol fuel credit and the biodiesel production tax credit, are discussed in Chapter 5. This chapter analyzes the impact of other energy-related tax expenditures, focusing on credits for electricity production from renewable resources (the renewable energy production and investment tax credits) and the depletion allowance tax preference (the tax provisions that allows capital costs for oil and gas wells to be recovered as a percent of revenues instead of costs).

The major provisions are analyzed using the National Energy Modeling System for the National Academy of Sciences (NEMS-NAS) model. The NEMS-NAS model was unable to capture other provisions—the special tax rate on reserve funds set aside by firms for the decommissioning of nuclear power plants and the credit for energy-efficiency improvements to existing homes—and these are discussed qualitatively.

Some introductory remarks on the impacts of the provisions will set the stage. Each of the provisions discussed in this chapter has only a small impact on the greenhouse gases emitted in the United States. This is to be expected given the nature and magnitude of the tax expenditures. Energy-related tax expenditures alter either the supply of or the demand for various types of energy. In some cases, the magnitude of the shift in supply or demand is small; hence, so is the estimated impact on energy consumption and CO_2 emissions. For example, the excess of percentage over cost depletion for natural gas lowers the cost of producing natural gas; however, the provision affects only independent producers, so the impact on natural gas production is small.

In the case of the renewable energy production tax credit, the credit, on a per-unit basis, is substantial: It lowers the cost of electricity production from wind by 2.3 cents per kWh, or almost 20 percent of the average retail price. However, the base to which the credit is applied (i.e., the fraction of electricity generated from wind) is small and so therefore is the impact on CO_2 emissions.[1]

The magnitude of estimated impacts also depends on the price responsiveness of energy consumption and production (technically, the price elasticities of supply and of demand), which depends on many factors, including the time horizon considered. Generally speaking, the price elasticity of demand for energy is higher in the long run than in the short run, since users are more easily able to adjust stocks of energy-using capital (e.g., appliances) when given longer time horizons. The same is true of the elasticity of energy supply: The price elasticity of supply of electricity from a particular fuel source will, in general, increase with the length of the time horizon considered. It is also true that the estimated impact of a tax credit will, in most cases, be greater the larger the price elasticities of demand and supply are for the affected energy source. Thus, long-run impacts are likely to differ from short-run impacts.

---

[1]Only 2.9 percent of electricity in the United States was generated from wind power in 2011.

FINDINGS FROM PRIOR LITERATURE

Several scholars and researchers have investigated the impact of the production tax credit on new installation of renewable electricity generation capacity. Some of those studies (Wiser, 2007; Metcalf, 2010) found that the tax credit for production of electricity from certain renewable resources (commonly called the Production Tax Credit, or PTC) did reduce the cost of installing new renewable generating capacity, especially for wind, but was costly to the Treasury Department (Metcalf, 2007). Those studies concluded that the PTC had increased the amount of installed generation capacity. One study (Price, 2002) found that the credit had no significant impact on new installations when state renewable portfolio standards were taken into account. Regardless of the findings on capacity, none of these studies estimated the production tax credit's impact on greenhouse gas emissions.

For other provisions discussed in this chapter, the existing literature was even thinner. Papers considering energy-efficiency improvement to existing homes looked at the possibility that homeowners taking the credit would have made the improvements in the absence of the incentive but did not evaluate any anticipated effect on emissions (Hirst et al., 1982; Metcalf, 2007; Jaffe and Stavins, 1994). Little has been written on the special tax rate on nuclear decommissioning reserve funds. A few studies have considered the impact of the percentage depletion allowance rules on levels of investment by firms in oil and gas wells (Krueger, 2009, and Metcalf, 2009). One other study estimated the impact of this subsidy on global petroleum production, while other researchers argue that global markets would largely offset any changes in U.S. production (Metcalf, 2007, and Bogdanski, 2011). None of these studies explicitly considered the impact of these provisions on GHG emissions.

ANALYSIS USING NEMS

The committee considered several modeling approaches to estimating the impact of the energy-sector tax provisions. As was explained in Chapter 2, the primary analysis was conducted using a version of the U.S. Energy Information Administration's (EIA) National Energy Modeling System, or NEMS, for the committee, with modifications made by a firm that maintains the model, OnLocation, Inc. This modified model is labeled the NEMS-NAS model.[2]

---

[2]The version of NEMS used in this study was run by OnLocation, Inc., for the National Academies and was run without a link to a macroeconomic model. The committee omitted the macroeconomic linkage because it included business-cycle linkages that were thought inappropriate for the long-run analysis undertaken here. Some model modifications were made in order to represent the NAS tax policy cases. To distinguish these re-

The Reference Scenario was developed starting with the Energy Information Administration's Annual Energy Outlook (AEO) for 2011 (U.S. Energy Information Administration, 2011) as the benchmark for the U.S. energy system (see Table 3-1). As with the other modeling efforts reported in later chapters, we standardized the modeling runs by assuming that the provisions of the Internal

TABLE 3-1 Assumptions Underlying NEMS Scenarios

	Reference (AEO 2011)	High Macro	High Oil Price	Low Gas Prices	No RPS
GDP Growth (real annual)	2.7%	3.2%	2.7%	2.7%	2.7%
2035 World Oil Price (2011 USD)	$125/bbl	$125/bbl	$200/bbl	$125/bbl	$125/bbl
U.S. Shale Gas Reserves				50% higher	
Renewable Portfolio Standards	Yes	Yes	Yes	Yes	No

Revenue Code (IRC) as of 2011 would remain in force through 2035 in the baseline (Reference) scenario. A set of counterfactual scenarios were then run, each scenario removing a particular tax provision beginning in 2010. By comparing the counterfactual (no-tax-preference) scenario to the reference (the baseline with-tax-preference) scenario, we were able to estimate the impact of the provision on GHG emissions and other related energy system and economic variables.

Note that for these simulations, as in all partial equilibrium models in this study, the economic impacts of the government's revenue gains or losses from changing a provision was omitted from the calculations. We do consider the impacts of recycling the gained or lost revenues in Chapter 6 when we examine the results for broad-based provisions studied with a general equilibrium model.

One of the issues arising in conducting modeling calculations of the kind reported here is to understand the uncertainties associated with the results. As with other models in this report, a formal quantification of the key uncertainties in NEMS was not conducted for this study. Based on its review and understanding of energy modeling, the committee determined that a full uncertainty study was not feasible within the constraints and resources available. However, for the NEMS model, sensitivity analyses were conducted to determine the impact of alternative assumptions on the results.

In AEO 2011, gross domestic product (GDP) is assumed to grow at an annual rate of 2.7 percent, and the price of oil in 2035 is assumed to be $125 per

---

sults from EIA results, this version of NEMS is called NEMS-NAS. For more detailed results and further materials, see the online Appendix to this report.

barrel (2011 USD). To examine how sensitive the results are to plausible alternative assumptions, three additional economic scenarios were run. These examined how alternative market conditions might affect the impacts of the tax provisions relative to the Reference scenario. We label these the High-Macroeconomic-Growth scenario, the High-Oil-Price scenario, and the Low-Gas-Prices scenario.

- The High-Macroeconomic-Growth scenario was run assuming a real GDP growth rate for the United States of 3.2 percent per year.

- The High-Oil-Price scenario assumed a 2035 price of $200 per barrel (2011 USD).
- The Low-Gas-Prices scenario assumed 50 percent higher ultimate recovery of natural gas from shale relative to the Reference scenario.

The inclusion of the High-Oil-Price and the Low-Gas-Price scenarios helped capture the impact of some of the major shifts in energy production and supply in the United States and then evaluated these changes against the influence of the tax provisions studied. For example, cumulative CO_2 emissions between 2015 and 2035 in the U.S. energy sector are projected to be 141,201 MMT in the Reference scenario. The High-Macroeconomic-Growth scenario projects cumulative emissions of 147,675 MMT, or about 4.5 percent greater than the Reference scenario. The High-Oil-Price scenario projection is 138,695 MMT, or about 6 percent lower, and the Low-Gas-Prices scenario projection is 140,616 MMT, or about 1.5 percent higher. A comparison of these changes in emissions can be seen in Tables 3-2a and b, which summarize the impacts of the selected provisions.

All four scenarios assumed that nonrenewable federal tax incentives and state Renewable Portfolio Standards (RPS) remain in place. The assumptions about environmental regulations are complicated in both the modeling and in reality. NEMS-NAS assumes that the Clean Air Interstate Rule (CAIR) will be implemented; however, it assumes that two other important rules are not implemented (the Cross-State Air Pollution Rule, CSAPR, and the rule on Mercury and Air Toxics Standards, MATS).

The legal status of these rules is somewhat different from the NEMS-NAS assumptions as of the time this report was completed. The MATS and CAIR are currently in force. The CSAPR is in a complicated legal limbo. The D.C. Circuit Court of Appeals vacated CSAPR in August 2012 and reinstated CAIR. However, in April 2013, the U.S. Environmental Protection Agency (EPA) asked the U.S. Supreme Court to review that decision. As a result, the lower court's ruling leaves CAIR in place until the litigation is resolved. The major implication for the present study is that these regulations might have a major impact on future electrical generation from coal. This, in turn, could influence the impact of tax provisions that affect the power sector.

In addition to the four aforementioned scenarios, we ran a No-RPS scenario. This was identical to the Reference scenario, but assumed that state Renewable Portfolio Standards would not remain in force over the modeling period. RPS require roughly 420 billion kWh of qualified renewable generation by 2035; hence, removing these requirements is likely to affect the impact of tax expenditures designed to promote renewable energy.

When analyzing the impact of removing the production tax credit/ investment tax credit (PTC/ITC), we ran all but the High-Oil-Price scenario. In analyzing the effects of replacing the percentage depletion allowance with cost depletion, we ran all scenarios except the No-RPS scenario.

NEMS-NAS RESULTS FOR THE PRODUCTION TAX CREDIT AND INVESTMENT TAX CREDIT

A summary of CO_2 emissions impacts from two main provisions analyzed with the NEMS-NAS model is presented in Table 3-2a. Cumulative and annual average CO_2 emissions in MMT from the U.S. energy sector are provided for the period 2010 to 2035 across the Reference scenario and the additional four sensitivity scenarios. For comparison, the impacts of removing the renewable energy production and investment tax credits and excess of percentage over cost depletion provisions are also provided for each scenario.

These calculations using the NEMS-NAS model indicate that the impacts of changes in these two tax provisions on CO_2 emissions are small across a spectrum of alternative market and regulatory conditions. The impact across policies and economic scenarios is between -0.1 and +0.3 percent of cumulative emissions from the energy sector over the period 2010-2035.

Results for the special Reference No-RPS scenario are included in Table 3-2b. This is the case where state RPS are not included in the Reference scenario and the PTC/ITCs are also removed. For comparison, the main Reference scenario emissions are also included. NEMS-NAS model results indicate a greater impact of removing the PTC/ITCs in the situation when there are no state RPS. For the No-RPS scenario, there is an increase of 0.5 percent in both cumulative and average annual emissions from the energy sector over the period 2010-2035. We discuss the finding in more detail in the following sections.

RENEWABLE ELECTRICITY TAX CREDITS (ENERGY PRODUCTION AND INVESTMENT TAX CREDITS)

Legal Description and Expected Impact

At the time of our analysis, taxpayers could claim a nonrefundable credit of 2.3 cents per kWh of electricity generated from wind, biomass, and geothermal energy resources and a credit of 1.1 cents per kWh for electricity generated from solar energy, small irrigation power, and municipal solid waste (trash

TABLE 3-2a Summary of CO_2 Emissions Impacts

U.S. Energy Sector CO_2 Emissions (MMT)	Reference Scenario (Base Value)	Difference from Respective Tax Policy Scenario		Percent Difference	
		No PTC/ITC	No Percentage Depletion	No PTC/ITC	No Percentage Depletion
Cumulative 2010-2035					
Reference Scenario	**141,201**	**360**	**-37**	**0.3%**	**-0.03%**
High Economic Growth	147,675	393	58	0.3%	0.04%
High Oil Prices	138,695	Nc	286	nc	0.2%
Low Gas Prices	140,616	-129	11	-0.1%	0.0%
Average Annual 2010-2035					
Reference Scenario	**5,883**	**15**	**-1.5**	**0.3%**	**-0.03%**
High Economic Growth	6,153	16	2.4	0.3%	0.04%
High Oil Prices	5,779	Nc	12	nc	0.2%
Low Gas Prices	5,859	-5.4	-0.5	-0.1%	0.0%

Source: NEMS-NAS model for this study. Note: "nc" is not calculated; MMT = million metric tons of CO_2.

TABLE 3-2b Summary of CO$_2$ Emissions Impacts

U.S. Energy Sector CO$_2$ Emissions (MMT)	Reference Scenario (Base Value)	Reference No-RPS Scenario	Difference from Reference No-RPS Scenario No PTC/ITC	Percent Difference No PTC/ITC
Cumulative 2010-2035				
Reference Scenario	141,201	141,576	762	0.5%
Average Annual 2010-2035				
Reference Scenario	5,883	5,899	32	0.5%

Source: NEMS-NAS model results for this study. Note: MMT = million metric tons of CO$_2$.

combustion and landfill gas) for the first 10 years after a facility is built.[3] This provision is commonly known as the Production Tax Credit (PTC). The American Recovery and Reinvestment Act (ARRA) of 2009 (P.L. 111-5) temporarily made the subsidy available as a cash grant in lieu of the tax credit, thereby making it refundable, that is, available to firms with no tax liability. The ARRA also enabled firms to take a 30 percent investment tax credit (ITC) in lieu of the 10-year PTC. Roughly 75 percent of all credits have gone to wind generation facilities and 16 percent to biomass facilities. In 2010, the Treasury Department estimated that the PTC/ITC reduced government revenues by $4.2 billion, comprising 36 percent of the estimated aggregate energy-related tax expenditures.[4]

The PTC and ITC lower the cost of electricity generated from renewable resources, encouraging increased substitution of renewable resources for coal or other fuels for electricity generation. They also lower the price of electricity and thereby increase overall demand. In many states, RPS also encourage electricity generation from renewable sources, and the tax credits reduce the cost of complying with the RPS.

The amount of GHG reduction from renewable energy generation depends on the source of the electricity. According to the EPA, the national average carbon dioxide output rate for electricity generated in 2009 was 1.2 lb CO_2 per kWh for delivered electricity.[5] To the extent that the PTC/ITC encourages the substitution of electricity from wind or solar power for electricity from fossil fuels, CO_2 emissions are expected to decrease.

Modeling in NEMS

We present the results of this scenario in detail. This will help readers understand the logic of the modeling results as well as give a glimpse into the complexity of the energy system and the difficulty of accurately capturing all the forces at work.

In each scenario, the baseline includes several production tax credits and investment tax credits related to renewable power generation, extending to the end of the NEMS-NAS forecast period (i.e., 2035). For the counterfactual No-

---

[3]The legislation, IRC section 45, set the credit at 1.5 cents per kWh of electricity generated from wind, biomass, and geothermalenergy resources and half that for electricity generated from solar energy, small irrigation power, and municipal solid waste (trash combustion and landfill gas). The credit is refundable and indexed to inflation. At the time of the analysis the inflation-indexed rate was 2.2 cents per kWh. At the time this report went to press, June 2013, the rate had risen to 2.3 cents per kWh.

[4]Percentage computation from estimates in Analytical Perspectives Table 17-1.

[5]EPA (U.S. Environmental Protection Agency). 2012. Emissions & Generation Resource Integrated Database (eGRID) 2012 Version 1.0, Year 2009 Summary Tables [online]. Available: http://www.epa.gov/cleanenergy/documents/egridzips/eGRID2012V1_0_year09_SummaryTables.pdf [accessed June 5, 2013].

PTC/ITC scenario, these tax credits were removed starting in 2010.[6] State renewable portfolio standards remain unchanged except in the No-RPS scenario. In all scenarios, nonrenewable tax credits remain unchanged, including credits for advanced coal, nuclear,[7] and combined heat and power.[8]

The ARRA enabled renewable project developers to choose a cash grant or a 30 percent ITC in lieu of a 10-year PTC. The NEMS-NAS Reference scenario assumes the following (values in 2009 USD, increasing with projected inflation):

- 2.3¢ per kWh PTC for onshore wind and geothermal;
- 1.1¢ per kWh PTC for landfill gas and hydroelectric facilities;
- 30 percent ITC per grant for biomass, offshore wind, and utility solar systems;
- 30 percent ITC per grant for distributed rooftop photovoltaic (PV) systems and small wind turbines.

In NEMS-NAS analysis, investment tax credits and cash grants are both treated as a percentage reduction in the capital cost of the technology and are therefore identical. Under current law, most of these provisions have expired or are scheduled to expire; however, under the committee's methodology, they are extended through 2035 in the baseline analysis for each scenario.

An important feature of the technological assumptions in the NEMS-NAS model is the introduction of learning by doing (see the discussion of this issue in Chapter 2).[9] The "learning rate" in the NEMS-NAS model is defined as the fractional reduction in capital costs for every doubling of cumulative capacity. The learning rates are determined separately for each component of the system. More specifically, each new technology is broken into its major components, and each component is identified as revolutionary, evolutionary, or mature. There is a minimum linear cost reduction for each component and also a formula for cost reductions based on new capacity additions. The resulting learning factor is based on whichever is greatest. The learning rates for onshore wind power, landfill gas, and hydropower are the same as for conventional fossil-fuel power gen-

---

[6]Facilities claiming the PTC beginning before 2012 are assumed to receive the credit for a 10-year period in the No-PTC/ITC scenario.

[7]The Energy Policy Act of 2005 provides a 20 percent investment tax credit for Integrated Coal Gasification Combined Cycle capacity and a 15 percent investment tax credit for other advanced coal technologies. Both of these are limited to 3 GW. There is also a production tax credit of 1.8 cents per kWh for new nuclear capacity beginning operation by 2020.

[8]The Energy Improvement and Extension Act of 2008 provides for a 10 percent tax credit for combined heat and power projects, applicable to only the first 15 MW of a system smaller than 50 MW. The system must be placed in service between October 2008 and December 2016.

[9]For full details and specifications, see Electricity Market Module, AEO Assumptions at http://www.eia.gov/forecasts/aeo/assumptions/pdf/electricity.pdf.

eration (coal and natural gas). The learning rates for advanced technologies, both renewables (offshore and solar PV) and fossil (carbon capture and storage), are higher but similar because they are not mature technologies. It should be noted that the assumptions about learning are difficult to establish empirically. Other models treat technological change in different ways.

Three sensitivity scenarios were examined for the PTC/ITC tax provision: High-Macroeconomic-Growth, Low-Natural-Gas-Prices, and No-RPS. The High-Macroeconomic-Growth scenario is expected to magnify the impact of the tax provisions, because this scenario increases the need for investments to meet higher levels of electricity demand. The Low-Natural-Gas-Prices scenario is expected to have its greatest impact in the electricity sector by reducing the cost-effectiveness of renewable generation.

Modeling Results

U. S. Electricity Generation

Compared to the Reference scenario, there is a slight increase in total electricity generation if the PTC/ITC tax preferences are removed (see Figure 3-1).

Generation from both coal and natural gas increases due to the removal of the tax credits, while renewable generation, especially end-use solar PV, decreases compared to the Reference scenario. Increased generation from biomass (mainly co-firing with coal) partially offsets decreases in wind and end-use renewable generation, as do new gas technologies. Biomass use increases primarily to help meet the requirements of state renewable electricity standards, as other eligible renewable generation declines (see Figure 3-2).

Changes in the Fuel Mix for Electric Power Generating Capacity

Comparing the Reference scenario baseline with the No-PTC/ITC scenario shows changes in generating-capacity additions and retirements. For example, NEMS-NAS projects that if the PTC/ITC are eliminated, then by 2035 utilities will add more than twice as many combustion turbines and nearly 50 percent more natural gas combined cycle plants while retiring 25 percent fewer coal-fired plants (compared to baseline projections where the PTC/ITC are still available). Compared to the Reference scenario, the No-PTC/ITC scenario projects about half as many new utility-scale renewable installations and one-fifth as much new end-use renewables. Total renewable capacity growth through 2035 is nearly 70 GW lower than NEMS-NAS predicts when the PTC/ITC is in place. These changes are shown in Figure 3-3.

The largest reduction is in rooftop photovoltaic systems (PVS). EIA's Reference scenario projects PVS will increase dramatically over time as their cost declines due to learning by doing (see the discussion about learning, above).

Because of learning, as more PVS are installed, the cost of installing systems is projected to decline. In the NEMS-NAS model this will lead to lower prices and higher market demand. The NEMS-NAS model assumes a fast learning rate for rooftop PVS, meaning that a small investment leads to rapid cost decreases and increasing installation rates. Removing the tax credits thus results in a decline in PVS investments. Because the EIA's Reference scenario predicts such rapid growth from learning by doing, the decline from removing the tax preference is large for PV.[10] Utility-scale wind deployment is also reduced by more than 15 GW by 2035. These changes, specific to non-hydro renewable generation, are shown in Figure 3-4.

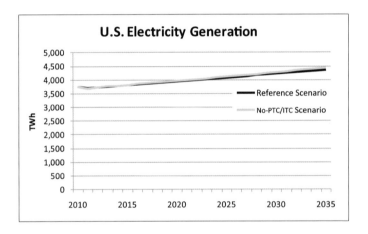

FIGURE 3-1 Total U. S. Electricity Generation, in Terawatt hours (TWh) – Reference Scenario and No-PTC/ITC Scenario.

---

[10]Additional details on NEMS-NAS model structure and features can be found on the NEMS Web site at www.eia.gov and in Appendix A of this report.

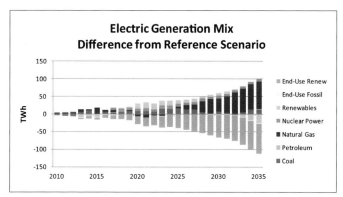

FIGURE 3-2 Changes in Electricity Generation Caused, in Terawatt hours (TWh), by Removing the PTC/ITC Compared to the Reference Scenario.

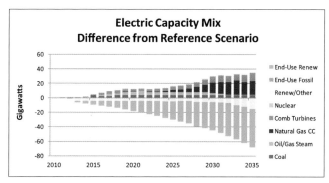

FIGURE 3-3 Changes in Electricity Generating Capacity Caused by Removing the PTC/ITC Compared to the Reference Scenario.

Electricity Prices

Removal of the renewable electricity credits raises the price of natural gas and electricity, which increases the cost of energy to consumers. When these credits are removed, the only remaining credits are the PTC payments for plants that came online prior to 2012 and are payable for the first 10 years of operation. All payments in this case end by 2021.

Compared to the Reference scenario, electricity prices increase by 0.2 cents per KWh or 1.8 percent by 2035 when tax credits are removed (see Figure 3-5). Investment tax credits help reduce utility costs for building new generation, and production tax credits reduce the cost of generation, both of which contribute to lower electricity rates for consumers in the Reference scenario baseline. These benefits disappear when the PTC/ITC credits are removed.

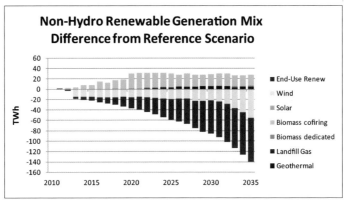

FIGURE 3-4 Changes in Non-Hydro Renewable Generation Caused by Removing the PTC/ITC Compared to the Reference Scenario.

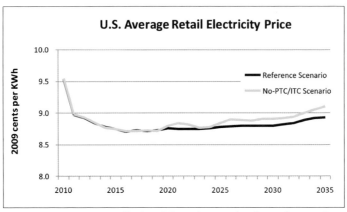

FIGURE 3-5 U.S. Average Retail Electricity Prices under the Reference Scenario and the No-PTC/ITC Scenario.

The impacts of removing the PTC/ITC on electricity demand in the NEMS-NAS model appear paradoxical. This results from the intricacies of the treatment of end-use (household and other) generation. Despite increased electricity prices in the No-PTC/ITC scenario, electricity sales from the grid increase due to the substitution of utility generation for end-use renewable generation. Households who would have installed rooftop PVS using the extended tax credits find it cheaper to purchase power from their local utility in the absence of these credits. End-use generation from renewables is also reduced, which increases primarily residential expenditures on electricity.

Sensitivity Analyses

The sensitivity analyses provide a wealth of detail, and a discussion is limited to a few highlights. Assuming a higher real annual GDP growth of 3.2 percent, the High-Macro-Growth scenario, NEMS-NAS predicts higher electricity demand when all of the tax credits are available. This implies that renewable electricity generation capacity is also projected to be higher. The impact of removing the PTC/ITC credits on renewable electricity generation is therefore magnified in the High-Macroeconomic-Growth scenario: There is greater reduction in electricity generation from renewables and a greater increase in the price of electricity when the credits are removed, compared to the Reference scenario GDP growth assumptions, as seen in Figures 3-2 and 3-5.

By contrast, in the Low-Natural-Gas-Prices scenario, natural gas generation is more cost-effective relative to renewable technologies. This implies that less renewable generation is built where all tax policies are in effect in the Low-Natural-Gas-Prices scenario than in the Reference scenario. Removing the PTC/ITC, therefore, has a smaller impact on the mix of generation capacity in the Low-Natural-Gas-Prices scenario than in the Reference scenario.

Tax Expenditures

Figure 3-6 shows projections of the impact of removing the PTC/ITC on government tax expenditures. In other words, these are the revenue losses for the different scenarios when the tax preferences are removed. Note that generation capacity that began receiving the PTC in 2012 is assumed to continue receiving the credit for the duration stipulated in the statute under the No-PTC/ITC Scenario.

The government's revenue losses on renewable electric generation facilities are between $4 and 5 billion per year in the Reference scenario. More than half of the expenditures, between $2 and 3 billion per year, come from investment tax credits to end-use renewable installations. Tax expenditures are even higher in the High-Macroeconomic-Growth scenario. In the No PTC/ITC scenario, the only remaining credits are the PTC payments for plants (primarily wind turbines) that came online prior to 2012 and are payable for the first 10 years of operation. All payments end by 2021 when the tax preferences are re-moved.

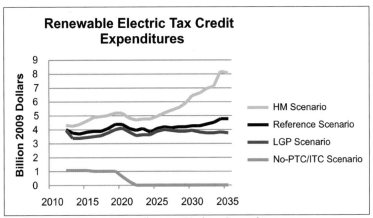

FIGURE 3-6 PTC/ITC Tax Expenditures – Various Scenarios.

CO₂ Emissions

Given the small changes in generation in the No-ITC/PTC scenario, changes to overall emissions from the domestic electric power sector also are small. Compared to the Reference scenario, removing the renewable electricity tax credits change CO_2 emissions on average by 15 MMT per year, or about 0.3 percent of power-sector emissions. The impact on CO_2 emissions, however, increases over time, reflecting the change in generation capacity (see Figure 3-7). Removing the electricity tax credits increases CO_2 emissions from the power sector by 42 MMT per year (or almost 2 percent of power-sector emissions) by 2035. Emissions from both natural gas and coal increase as generation from these fuels increases.

The impact on CO_2 emissions of removing the PTC/ITC is larger in the High-Macroeconomic-Growth scenario: The increase in CO_2 emissions over the period 2031-2035 is twice as large in the High-Macro-Growth scenario as in the Reference scenario, increasing CO_2 emissions from the power sector by 0.8 percent.

In the Reference and High-Macroeconomic-Growth scenarios, utilities increasingly rely on natural gas for electricity generation, including construction of new generating capacity. Gas replaces most of the reduced renewable generation in the No-ITC/PTC scenario, and coal and nuclear power contribute modestly in some instances. In the Low-Gas-Price scenario, more nuclear power plants that were on the cusp of retirement in the Reference scenario remain in use, and as a result the increase in CO_2 emissions is close to the Reference scenario.

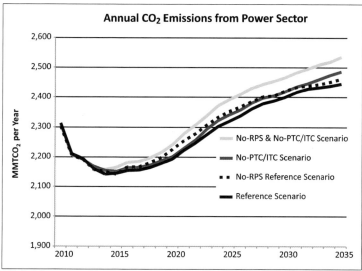

FIGURE 3-7 Power sector CO_2 emission scenarios showing effect of renewable portfolio standards.

The Role of Renewable Portfolio Standards

As a final sensitivity run, we examined the impacts of changing tax provisions if state Renewable Portfolio Standards are removed. This scenario is important for illustrating the interaction of regulatory mandates with tax policy. We ran four scenarios using the NEMS-NAS model. One set, discussed above, is with the RPS and with and without the ITC/PTC tax preferences, and then another set to calculate the model without the RPS and with and without the ITC/PTC tax preferences.

The results of these four scenarios are illustrated in Figure 3-7 and can be summarized as follows. To begin with, the Reference scenario—with both the ITC/PTC and RPS policies—has the lowest power-sector CO_2 emissions trajectory out to 2035, with cumulative emissions of 54,879 MMT. Removing the ITC/PTC tax preference but retaining the RPS policies (RPS/No-PTC/ITC scenario) yields higher CO_2 emissions from the power sector, as expected, resulting in 55,255 MMT cumulative emissions between 2010 and 2035 (which is 0.7 percent higher).

To assess the impact of having the RPS in place, we did the PTC/ITC and the No- PTC/ITC simulations without the RPS. The No-RPS Reference scenario resulted in somewhat higher cumulative power-sector CO_2 emissions of 55,315

MMT (or 0.8 percent higher) than the Reference scenario. Thus, the RPS policies have, according to the NEMS-NAS modeling, virtually the same effect in reducing GHG emissions as the PTC/ITC tax preferences.

Last, we considered the effect of removing the PTC/ITC tax preferences if the RPS mandates are not in place (No-RPS and No-PTC/ITC scenario). According to the simulation, cumulative power-sector CO_2 emissions are 56,124 MMT (or 1.5 percent) above the No-RPS/Reference scenario. In other words, the increase of CO_2 emissions, the result of removing the PTC/ITC tax credits, are about twice as large if RPS mandates are not in effect.

The finding on the role of the RPS is important. It indicates that the regulatory mandates constrain production and emissions. As a result, the impacts of tax policies on emissions are reduced, in this case by half, when the regulatory mandates are considered. This finding is similar in other results, particularly the impacts of the biofuels mandates analyzed in Chapter 5. While the exact magnitudes will be sensitive to the detailed specifications, the general point about including regulations and mandates in estimating the impacts of tax policy should be emphasized.

Summary

The committee's analysis of the tax provisions for renewable electricity indicates that they lower CO_2 emissions. This finding confirms the first-order intuition that lowering the cost of low-carbon renewable fuels will lead to substitution away from high-carbon fossil fuels.

The reduction in CO_2 emissions associated with the PTC/ITC is, however, small, amounting to about 0.3 percent of CO_2 emissions from the energy sector in the Reference scenario. If the revenue lost as a result of the PTC/ITC is divided by the reduction in CO_2 emissions, just under $250 in revenues are lost per ton of CO_2 reduced. While this does not represent the social cost of reducing the ton of CO_2 emissions (because revenue losses are not a dead-weight loss, as explained in Chapter 2), the fiscal cost per ton of CO_2 reduced is high relative to other, more efficient approaches.

EXCESS OF PERCENTAGE OVER COST DEPLETION

Legal Description and Expected Impact

The second provision described in detail is the depletion allowance. The depletion allowance permits owners of oil and gas wells to deduct the decline in the value of their reserves as oil or gas is extracted and sold. The allowance, which is a form of cost recovery for capital investments, can be calculated using either cost depletion or percentage depletion. Under cost depletion, the annual deduction is equal to the unrecovered cost of acquisition and development of the resource times the estimated proportion of the resource removed during that year. Under percentage depletion, taxpayers deduct a percentage of gross income associated with the sale of the resource. Percentage depletion for oil and gas is currently limited to U.S. production by independent companies up to a certain limit, currently set at 15 percent of costs associated with production. Percentage depletion typically allows for total deductions that exceed the cost of capital invested to acquire and develop the resource. A percentage depletion rate of 22 percent applies to natural gas sold under fixed contracts.

The excess of percentage over cost depletion affects primarily natural gas production. Since only independent producers may utilize this tax subsidy, the Joint Committee on Taxation (JCT) estimates its total impact to the Treasury at $0.5 billion for 2010. With increasing production, it reaches $1.0 billion in 2014, with a 5-year total of $4.1 billion.

For natural gas development and production, the largest component of capital investment is the drilling of exploration and production wells. Any policy affecting capital investment directly, as the depletion allowance does, affects this front-end activity in the natural gas market life cycle. In the Cost-Depletion scenario, where cost depletion replaces the percentage depletion allowance, capital recovery is slower, resulting in higher drilling costs, and reducing incentives to explore and develop new supply. Less investment in drilling would be expected to reduce domestic production and raise the price of natural gas.

Modeling in NEMS

Explicit treatment of tax deductions associated with resource depletion is limited in the NEMS model. The tax treatment of depletion in the coal-mining sector is not explicitly modeled in NEMS-NAS and is therefore excluded from the Reference and Cost-Depletion scenarios. Current estimates are that the depletion allowance for coal is very small, so this is unlikely to have a large impact on the results.[11] The treatment of depletion in the oil and gas sector is complicated by the fact that the depletion allowance depends on firm size (small independent, large independent, and major producer). To capture this:

- In the Reference scenario, a 15 percent depletion allowance is assumed for all onshore activity, all of which is assumed to be performed by independent oil and gas producers.
- All offshore gas and oil production is assumed to be owned by major companies, which are not allowed to use the percentage depletion allowance.
- A 22 percent depletion allowance (typically reserved for natural gas sold under fixed contracts) was run as a sensitivity case.

---

[11]The Comptroller estimates coal's share of the percentage depletion allowance to be $29.7 million in 2006. See http://www.window.state.tx.us/specialrpt/energy/subsidies/index.php#coal.

MODELING RESULTS

Natural Gas Production and Consumption

The modeling finds that removing the percentage depletion preference results in higher drilling costs, reducing incentives to explore and develop new supply. The reduced incentives to drill wells lead to a reduction in domestic natural gas production of 14 Tcf over the 2010–2035 timespan in the Reference scenario, as shown in Figure 3-8. A very modest increase in natural gas imports of 0.4 Tcf is not sufficient to offset the drop in domestic production.

Lower supply leads to an increase in gas prices, as shown in Figure 3-9. (The dashed black and green lines correspond to the Reference scenario and Cost-Depletion scenarios, respectively.) This, in turn, reduces natural gas consumption by 2 percent, on average.

A sensitivity scenario was run where a cost depletion allowance of 22 percent was applied to the same resources that were otherwise receiving the benefit of the 15 percent allowance, further lowering exploration and production costs. While the impact is directionally consistent (higher allowance leads to lower gas prices, higher production, and consumption), the impact was minor, changing gas prices by 1.2 percent and production and consumption by less than 0.5 percent.

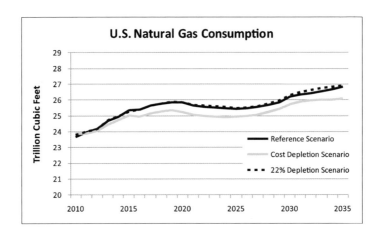

FIGURE 3-8 Natural Gas Consumption Under Three Scenarios.

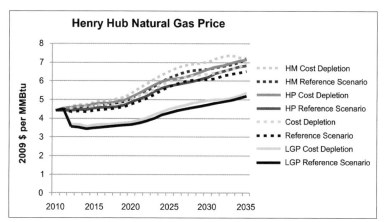

FIGURE 3-9 Natural Gas Prices Under Several Scenarios.

Electricity-Generation Mix

The primary impact of removing the percentage depletion tax preference is to increase the cost of natural gas production and, hence, natural gas prices. All sectors reduce their natural gas consumption. However, the biggest impact occurs in the power sector because substitution of other fuels is easiest there.

Depending on underlying market conditions, natural gas is replaced by coal, nuclear, and/or renewable generation. Of the renewable resources filling a portion of the generation gap created by lower levels of natural gas generation, biomass and wind contribute the greatest amount. Compared to the Reference scenario, wind generates 122 TWh more electricity, and biomass (co-firing) contributes an additional 63 TWh over the model time horizon.

Compared to the Reference scenario, there are changes in the pattern of retirement of nuclear plants because gas prices are higher. End-use fossil fuel electricity generation (primarily gas-fired combined heat and power) also declines.

Sensitivity Analysis

The impact of removing percentage depletion was examined for three alternative economic assumptions: high macroeconomic growth, high oil prices, and low natural gas prices.

The High-Macroeconomic-Growth scenario results in higher energy consumption and energy prices, which magnify the impact of the tax provision. The High-Oil-Price scenario increases the impact of replacing percentage with cost depletion. The Low-Natural-Gas-Prices scenario reduces the impact of the excess of percentage over cost depletion.

As intuition would suggest, the primary impact of the move to cost depletion from percentage depletion is to increase the cost of natural gas production and prices, with the High-Macroeconomic-Growth scenario showing the largest difference and the Low-Natural-Gas-Prices scenario showing the least difference.

All sectors reduce their natural gas consumption, with the biggest impact occurring in the electricity sector. Depending on underlying market conditions, the reduction in gas is replaced by coal, nuclear, and/or renewable generation.

In the Reference and Low-Natural-Gas-Prices scenarios, there are some nuclear plants whose retirements are postponed in the Cost-Depletion scenario. In addition, end-use electricity generation from gas-fired combined heat and power plants also declines.

CO_2 Emissions

The impact on CO_2 emissions of removing the percentage depletion allowance is small under all four scenarios. In the Reference scenario, there is a net reduction in CO_2 emissions of approximately 37 MMT over the model time horizon, summed across all sectors. This implies an average reduction of 1.5 MMT per year (see Table 3-2), or 0.03 percent of total CO_2 emissions. Higher gas prices, as a result of cost depletion, discourage generation from natural gas, which is replaced primarily by more carbon-intensive coal generation and by renewable generation. There is a slight increase in renewable generation to meet the gap caused by the shift away from natural gas, but this is not sufficient to offset the greater use of coal except after 2030. Figure 3-10 shows these projected changes in CO2 emissions.

In the other three scenarios there are small increases in CO_2 emissions when the percentage depletion allowance is removed. The Low-Natural-Gas-Prices scenario reduces the impact of the otherwise relatively higher natural gas prices. The High-Macroeconomic-Growth scenario leads to higher energy consumption and energy prices, which magnifies the fuel switching from natural gas to higher-carbon fuels and, hence, to slightly higher CO_2 emissions. Lastly, the High-Oil-Price scenario displays the greatest increase in emissions: 0.2 percent over the 2010-2035 period. In this scenario, the shift in the power sector is most dominant due to the loss of the relatively less expensive natural gas. It should, however, be emphasized that the magnitude of changes in CO_2 emissions is in all these scenarios extremely small, especially in the Low-Natural-Gas-Prices and Reference scenarios.

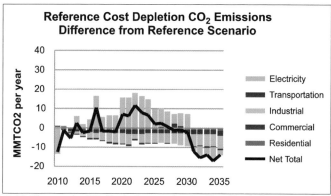

FIGURE 3-10 Changes in CO_2 Emissions under the Cost Depletion Scenario.

Summary

The primary impact of the depletion allowance is its impact on the production of natural gas and the spillover impacts in other markets. To a first approximation, the depletion allowance produces no impact on greenhouse gas emissions. While natural gas production goes down when percentage depletion is removed, the complex substitution patterns lead to largely offsetting forces and to a minimal overall impact on CO_2 and other GHG emissions. The four scenarios examined here have different signs in their impacts on CO_2 emissions, although each of them is tiny. We conclude that the sign of the change—whether it is positive or negative—is in reality uncertain but the size of the effect is likely to be very small.

From a fiscal point of view, the oil depletion allowance was not motivated by concerns about climate change when it was enacted in 1926. From the point of view of climate change, this is not an effective subsidy for reducing emissions.

CREDIT FOR ENERGY-EFFICIENCY
IMPROVEMENTS TO EXISTING HOMES

Legal Description

Homeowners can benefit from two tax credits for adding energy-efficiency improvements to their homes. The Qualified Energy Efficiency Improvements (IRC Section 25C) provides a 10 percent credit for the purchase of qualified energy-efficiency improvements to existing homes. The energy-efficiency home products must have been placed in service between January 1, 2011, and December 31, 2011. Under section 25C, the maximum credit for a taxpayer for all taxable years is $500, and no more than $200 of the credit may be attributable to expenditures on windows.

The Wind, Solar, Geothermal and Fuel Cell Tax Credit (IRC Section 25D) provides tax credits equal to 30 percent of the cost, with no cap through 2016, for construction of geothermal heat pumps, solar energy systems, solar water heaters, and small wind-energy systems and fuel cells.[12] The energy-efficiency products must be placed in service before the end of 2016. The credits are valid only for improvements made to the taxpayer's principal residence, except for qualified geothermal, solar, and wind property, which can be installed on any home used as a residence by the taxpayer.

Short Description of Economic-Fiscal Impact

The credit for energy-efficiency improvements to existing homes is meant to encourage the installation of energy-efficiency technologies in homes by decreasing the costs of installation. The purpose is to lower energy use in the residential sector, which should result in lower GHG emissions from energy consumption. The JCT estimated that these provisions would cause tax expenditures of $1.7 billion in fiscal year 2010 and expenditures of $2.9 billion from 2010–2014.

There is little solid empirical work on the impact of this provision on greenhouse gas emissions. The literature contains some theoretical and suggestive empirical evidence that points to positive impacts of government policies on energy-efficiency investment in homes. Theoretical models that consider various regulations, subsidies, or informational provision find that government intervention can help drive adoption of energy-saving technology. Both theoretical and empirical work suggests that market and behavioral failures (e.g., externalities, principal-agent issues, and informational barriers) can cause underinvestment in residential energy efficiency, and that government intervention can help.

A recurring theme in the literature is that energy prices to end-use consumers are distorted and are likely to understate the true marginal cost of energy. This suggests energy should be priced more accurately to reflect its true cost, which may encourage energy-efficient investments.

There is a literature that models empirically consumer purchases of energy-consuming appliances and the associated demand for energy; however, it is difficult to apply the results of this literature to the provisions in sections 25C and 25D of the tax code. The Center for Business and Economic Research (CBER) study[13] modeled estimates of the impact of these provisions and found that they led to reductions in CO_2 emissions of 4.2 MMT in 2009. However, the methodology used to derive this estimate has not been sufficiently validated empirically to allow the committee to adopt this estimate. The committee be-

---

[12]On fuel cells the credit cannot exceed $500 per 0.5 kW of installed capacity.

[13]University of Nevada, Las Vegas, Center for Business and Economic Research (CBER) paper by Allaire and Brown, 2011.

lieves that this is probably an upper bound on the impact, but the actual impact may be substantially smaller.

In the absence of detailed and reliable results in the existing literature, the committee investigated the possibility of undertaking modeling along the lines of the provisions discussed elsewhere in this report. Such modeling would require estimating the impact of credits on the prices to consumers of energy-consuming capital goods; then to calculate the impact of these price changes on investments made by homeowners; then to further calculate the impact of the changed energy-efficient investments on energy demand by fuel; and finally to calculate the impacts on greenhouse gas emissions.

It became apparent that none of the existing models was well designed to do this analysis. In particular, the NEMS-NAS model has no policy levers that translate tax changes into prices and then into energy demands. For this reason, the committee was unable to provide estimates that it found reliable for the impact of this provision.

Summary

The committee did not find, and was unable within its time and resource budget to produce, detailed and reliable estimates of the impact of the Credit for Energy Efficiency Improvements to Existing Homes. In practical terms, the effect of this section of the tax code on GHG emissions is likely to be limited. The reactions of households may be small because of (a) the volatility of the credits, which have varied in availability and in amount over time; (b) the relatively small credit limit (with the maximum being $500); and (c) the complexity of the provisions.

Notwithstanding these reservations, existing research points to the potential importance of incentives in this area. The combination of high potential payoff and limited research leads the committee to conclude that research on understanding the impacts of tax incentives on household energy consumption and GHG emissions should be encouraged.

SPECIAL TAX RATE ON NUCLEAR
DECOMMISSIONING RESERVE FUNDS

Legal Description

Nuclear power plant operators can elect to set aside reserve funds for the decommissioning of plants. The code provides for special tax treatment of these funds in two ways. Contributions are deductible in the year they are made and not taxed until distribution. This defers tax on those funds into the future. Once distributed, the funds, along with any gains from investments through the years, are taxed at a rate of 20 percent. Most utilities operating nuclear plants are large enough that their income would normally be taxed at 35 percent. Thus, a plant operator benefits from a deferred, lowered tax on the income made from the reserved funds.

Short Description of Economic-Fiscal Impact

When a nuclear power plant is retired from service (decommissioned), the residual radioactivity at the facility must be reduced to a level that allows transfer of the property. The U.S. Nuclear Regulatory Commission (USNRC) has rules governing nuclear power plant decommissioning. These involve cleanup of radioactively contaminated plant systems and structures and removal of the radioactive fuel. Before a nuclear power plant begins operations, the operator must ensure that there will be sufficient funds to cover the ultimate decommissioning of the facility. Each plant operator must report biannually to the USNRC the status of its decommissioning funds for each unit.[14]

According to the USNRC, the estimated decommissioning cost for a nuclear reactor can range from $300 to over $600 million, or between 10 and 25 percent of construction costs.[15] The total cost of decommissioning a nuclear reactor depends on the timing and sequence of the various stages of the construction program, the reactor type, the location of the facility, the radioactive-waste burial costs, and the ultimate plans for spent-fuel storage. Realized decommissioning costs have reached over $1 billion at some plants.

Given the size of decommissioning costs, a utility responsible for decommissioning a nuclear power plant can create a reserve fund to pay for these costs. About 70 percent of current operators are authorized to accumulate decommissioning funds over their plants' operational lifetimes.

Analysis and Summary

The committee was unable to find any detailed and reliable estimates of the impact of the nuclear decommissioning tax preference on greenhouse gas emissions.[16] This is a particularly difficult provision to analyze for several reasons. First, nuclear power plants have a very long useful lifespans (the lifetime of the plant could well be at least 60 years), and that time span extends beyond

---

[14]Currently, the U.S. power sector operates 104 commercial nuclear power plants. Most were built in the 1970s and are scheduled for decommissioning during the next three decades. As of April 2011, there were 23 nuclear units in various stages of decommissioning, with 10 of those completely cleaned up.

[15]For more information, see http://www.nrc.gov/reading-rm/doc-collections/fact-sheets/decommissioning.html.

[16]The CBER model did calculations of the impact of removing the nuclear decommissioning tax credit. However, it was in a supply-and-demand framework that was static and did not have the same detailed treatment of taxes, costs, and load curves as the NEMS-NAS model. The committee therefore finds that the CBER results are an upper bound of potential results, with the more likely lower bound being zero.

the time horizon of any of the models used for this study. Second, the provision involves complex computations of the opportunity cost of funds, the return on capital, as well as the future regulatory treatment of these costs under federal and state public utility laws and regulations. Third, the runs that were undertaken for the committee in the NEMS-NAS model actually project that there will be no new nuclear power plants licensed in the time period of the runs. Given that there are no new plants in the base run, this number clearly cannot be reduced by removing a tax preference. Therefore, to the extent that these results are considered reliable, the best estimate of the impact on nuclear power plant construction would be zero. There might, however, be an effect on extending the lifespan of existing plants.

While the committee was unable to find or commission detailed and reliable modeling runs on the nuclear decommissioning tax preference, it finds that the most likely impact is a negligible impact on greenhouse gas emissions.

SUMMARY

The present chapter examines some of the important energy-related tax expenditures. Two of them have been examined in detail using the NEMS-NAS model, while the other two were examined qualitatively. Several other provisions were not modeled, but the current four comprise more than half of all energy-related tax expenditures.

One important result is that the net estimated impact of the modeled provisions is very small. The central estimate of the net impact of the renewable tax credits, the depletion allowance, and the nuclear decommissioning credit is about 0.1 percent of total national GHG emissions over the next quarter-century. While the central estimate is close to zero, alternative assumptions about GDP growth, natural gas costs, and oil prices make the range larger and include both positive and negative numbers. We estimate that the net total is likely to be in the range of plus or minus ½ percent of total U.S. greenhouse gas emissions over this period.

A second result is that the web of interacting impacts is extremely complex and often leads to counterintuitive results. For example, the renewable electricity tax credit not only lowers costs but also apparently decreases total electricity output because of changes in the composition of generation. The depletion allowance, which is usually associated with oil, has its major effect on gas. The nuclear decommissioning credit is likely to have no effect at all because the number of nuclear power plants triggered by its existence is likely to be zero given that the number of new plants in most projections is already zero. Moreover, these estimates are undertaken in a partial equilibrium framework, and, as we will see in subsequent chapters, including the reaction of other sectors in a general equilibrium model makes the reactions even more complex and difficult to estimate accurately.

Third, the committee found that some of the provisions (such as the credit for energy-efficiency improvements to existing homes) are so complex, or involve so many unanswered questions, that it is not possible to provide reliable estimates of their impacts on greenhouse gas emissions. For both the impact of tax credits for energy efficiency in homes and the nuclear decommissioning tax preference, the uncertainties about future tax, regulatory, financial, and behavioral responses are so large that the committee was unable to provide what it regarded as reliable estimates. For other provisions, the estimates cannot resolve whether the net impact is negative or positive.

Finally, the energy-sector tax preferences are a good example of the reasons that tax specialists hold tax expenditures in low regard. Many of them were introduced in an earlier era and have outlived their original purposes. Others were introduced in order to foster important national goals, but the complexity of the energy-sector interactions actually leads to perverse impacts that are contrary to the original purposes.

Chapter 4

Energy-Related Excise Taxes

INTRODUCTION

All of the other analyses in this report consider the impact of subsidies from tax expenditures. The present chapter considers the impact of two energy-related excise taxes: highway motor fuels excise taxes and taxes on commercial aviation fuel. These taxes are important because the sectors to which they apply are highly GHG-intensive. Emissions from passenger vehicles, including cars and light trucks, account for more than 20 percent of U.S. GHG emissions, and airlines add approximately 2.5 percent of U.S. GHG emissions (U.S. Environmental Protection Agency, 2012).

The federal government taxes gasoline for on-road use at $0.184 per gallon and on-road diesel at $0.244 per gallon.[1] It is important that motor fuels are taxed on a volumetric basis rather than on the basis of energy content. The committee asked all four economic modeling contractors to estimate impacts on greenhouse gases of removing these federal highway motor fuels excise taxes. Comparing results across models provides a basis for understanding the mechanisms at work in the models as well as an estimate of the uncertainty of the estimates due to model differences. Summary results are presented in Table 4-1 below.

Several types of taxes are currently levied on U.S. domestic air travel, including the federal ticket tax, the flight segment fee, the passenger facility charge, and the federal security fee (Borenstein, 2011). In addition, commercial aviation pays an excise tax of $0.044 per gallon of jet fuel. Jet fuel for noncommercial aviation is taxed at a higher rate of $0.219 per gallon. According to Airlines for America, the total taxes levied on a typical $300 round-trip ticket is $61 (20 percent), but fuel taxes account for a very small share of the total tax

---

[1]Internal Revenue Code sections 4041 and 4083 impose a tax on fuels to pay into the Highway Trust Fund (authorized in IRC section 9503) and a tax to fund the Leaking Underground Storage Tank Trust Fund. The rates quoted here are the combination of both taxes.

burden (Airlines for America, 2012). Unfortunately, none of the models used for this study was able to analyze the federal ticket tax, the flight segment fee, the passenger facility charge, or the federal security fee. Moreover, only one model analyzed the excise tax on aviation fuel.

FINDINGS FROM PRIOR LITERATURE

Unlike the other provisions analyzed in this study, there is a large literature on the tax on highway motor fuels, particularly on gasoline. There are literally dozens of studies that have estimated the price-elasticity of demand—short run, long run, or both—for gasoline in the U.S. (Gillingham K. , 2011) Price elasticity of demand measures the sensitivity of the quantity demanded of a good when the price of the good changes, other things held constant. Though demand elasticities do not directly determine greenhouse gas impacts, they are an intermediate step to computing emissions. The committee uncovered 11 studies that quantify the impacts of price on vehicle miles traveled. These studies all find that, as the cost of driving increases, vehicles are driven less, as one would expect. When vehicles are driven less, consumption of motor fuels and GHG emissions are expected to decrease.

Another vein in the literature finds that increasing fuel costs can accelerate turnover in vehicle stock, influencing consumers to purchase more fuel efficient vehicles when they purchase new vehicles (Busse et al. 2013). More fuel efficient vehicles are expected to lead to lower emissions per quantity of fuel consumed, though not necessarily less fuel consumed if the rebound effect is large enough. (The rebound effect in this context refers to the tendency of drivers to increase the number of miles driven when the cost per mile driven declines as cars become more fuel efficient. (Greening et al., 2000 and Small and Dender, 2007).

In contrast to the rich literature that we reviewed on the excise tax on motor vehicle fuel, our literature review did not uncover any studies of the taxes on aviation fuel or passengers airline tickets.

ANALYSIS OF EXCISE TAXES ON HIGHWAY FUELS

The Internal Revenue Code levies a tax of $0.184 per gallon of gasoline or alcohol fuel for on-road use. This tax applies whether the fuels are pure or blended, such as the common gasoline plus ethanol mixtures.[2] The Code also levies on a tax on diesel fuels at $0.244 per gallon for diesel and kerosene and $0.197 per gallon for diesel-water fuel emulsion.[3]

---

[2]See Internal Revenue Code section IRC 4081.
[3]See Internal Revenue Code section IRC 4081.

Unlike the other provisions we considered, highway fuels excise taxes are a revenue source rather than a tax expenditure. The Internal Revenue Service's Statistics of Income division collects and reports information on excise tax receipts. In 2010, federal excise taxes on gasoline totaled approximately $25 billion, or 53 percent of all federal excise tax receipts. On-road diesel accounted for an additional $8.6 billion. Combined highway fuels taxes account for $33.7 billion, or over 71 percent of all federal excise taxes (U.S. Internal Revenue Service, 2012).

We also note that the impacts of the highway motor fuels taxes interact with other taxes and regulations in this sector. In particular, the tax credits for biofuels and the renewable fuels standards (RFS) will be major constraints on the effects of the highway taxes. A fuller discussion of these is contained in Chapter 5.

Modeling Results

The committee asked the four modeling teams that were engaged to estimate the greenhouse gas impacts of removing the highway motor fuels excise taxes. Summary results are presented in Table 4-1. All four models find that GHG emissions will increase if the taxes are removed. However, the estimates differ greatly across different models. Variations in the results across the models highlight differences in both the modeling approaches and assumptions used.

TABLE 4-1 Summary Impacts of Removing Federal Highway Fuels Taxes Across Four Models

	IGEM	NEMS	CBER	FAPRI
Assumption				
Modeling Period	2010-2035	2010-2035	2010-2035	2014-2021
Real GDP Growth	2.7%	2.6%	2.6%	2.6%
Energy Cons. Growth	0.48%	0.4%	0.4%	0.4%
GHG Included	$CO_2,CH_4,$ $N_2O,$HGWP	CO_2	CO_2	CO_2,CH_4, N_2O
Regional coverage	US	US	World (ltd)	World
Result: Increase in[a]				
Cumulative CO_2 emissions (MMT)	2,158	88	1,400	78.6
Avg. annual CO_2-e emissions (MMT/year)	83	3.5	54	9.8
Cumulative CO_2-e emissions (MMT)	2,173	NA	NA	NA

[a]Note that NEMS and CBER include only CO_2 emissions, while the other models include other non-CO_2 greenhouse-gas emissions.

Discussion of Modeling Results on Highway Fuels Excise Taxes

It will be useful to begin with some general discussion about the models. NEMS, CBER and FAPRI are all partial equilibrium models. This means they describe aspects of the modeled sector—in this case use of liquid fuels—with considerable detail. They do not, however, reflect the interactions between the energy sector and other parts of the economy. For instance, in the highway motor fuels excise taxes scenarios, NEMS, CBER and FAPRI do not model the impact of the increased federal revenues used to replace lost excise tax receipts on the composition of output and do not model emissions from increased consumer spending on non-energy goods and services. The results therefore primarily reflect the first-order response of consumers and producers to lower transportation fuel expenses, which in turn reflect embedded price elasticities of demand and supply for transportation fuels.

There are also differences across the three partial equilibrium models. NEMS has more fine-grained details on many aspects of the energy sector but focuses on the U.S. alone, while FAPRI and CBER model aspects of world energy markets. While FAPRI and CBER allow the world oil price to change in response to U.S. consumption, NEMS holds the world oil price fixed. FAPRI was designed with particular attention to biofuels, so it has more detail on the agricultural sector than NEMS or CBER. CBER is a more stylized model than either FAPRI or NEMS. This makes it less realistic but more transparent.

Additionally, the NEMS-NAS and CBER models include only CO_2 emissions, while the other models include other non-CO_2 greenhouse gases. This omission is very small in the models that include non-CO_2 emissions.

One advantage of the partial equilibrium models is that they represent specific components of the energy sector. For instance, both NEMS and FAPRI distinguish between taxes on gasoline and taxes on ethanol, and both sets of results suggest this is important in modeling the removal of the highway fuels taxes. The taxes on highway fuels are assessed per gallon, and ethanol has lower energy content per gallon than gasoline. This means that when the highway taxes are removed, the price per unit energy of ethanol declines by more in percentage terms than the price per unit energy of gasoline. By contrast, CBER does not distinguish between gasoline and ethanol, which explains in part why the CBER estimates are larger than the other two partial equilibrium models. In addition, CBER assumes more price-elastic demand for highway fuels than the other two models.

Given that the results from the partial equilibrium models mainly reflect highway fuels' demand and supply elasticities, it is useful to consider what behavior and decisions the elasticities capture (see Gillingham K. T., 2011 for more detail). Economists usually distinguish between long- and short-run price elasticities. For instance, in the case of gasoline, the short-run price elasticity reflects adjustments that consumers and businesses make that do not involve adjusting the type of vehicles they own. Consumers may drive fewer miles, perhaps because they opt not to take discretionary trips, they carpool to work or they take public transportation. Consumers may also adjust their driving habits, for instance, driving more slowly on highways, which would reduce fuel use per mile traveled. Or, households with multiple cars may shift their usage to their more fuel efficient car. Long-run price elasticities include adjustments to the stock of vehicles. If fuel prices are lower, as in the scenarios where federal highway fuels excise taxes are removed, consumers may purchase less fuel-efficient cars and may retire their less fuel-efficient cars more slowly. Conceptually, IGEM and CBER apply long-run elasticities to do their calculations, while NEMS-NAS and FAPRI have time-varying elasticities.

Appraisal of Individual Model Results

We now discuss and assess the individual model results. We begin with the IGEM results. IGEM has estimates of the impacts of removing the highway fuels excise taxes that are larger than those of the other studies. Having reviewed the calculations, the committee concludes that the IGEM model cannot accurately capture the structure of the motor fuels tax provisions and the associated regulations and therefore cannot provide reliable results for these provisions. IGEM does not contain a detailed sectoral description of the transportation sector. It does not have a detailed treatment of gasoline or highway fuels, of the properties of vehicles, of vehicle-miles travelled, or of the substitute fuels. It does not reflect the different energy content per gallon of different fuels. The IGEM experiment changed taxes on refined petroleum products, not on highway fuels. Finally, IGEM does not include the renewable fuel standards (RFS), so it cannot capture the RFS's constraints on inter-fuel substitution and on the mix between gasoline and ethanol. The strengths of IGEM – the capabilities to capture the impacts of the rest of the economy and the recycling of revenues – cannot offset its shortcomings in analyzing the effects of highway fuels taxes.

The second approach – the CBER model – has the advantage of transparency and reliance on estimates in the literature for its price-elasticities. However, in this context, it has three important shortcomings. First, it is a static model, and its elasticities are long-run rather than short-run. While the extent of the overestimate will depend upon the dynamics, it is likely that this would lead to an overestimate of the response by a substantial margin. Second, as will be discussed for the next two models, the CBER model does not contain a realistic representation of the renewable fuel standards, which are likely to constrain production and reduce the impact of taxes on emissions. Third, the price-elasticity of demand for petroleum in the transportation sector is assumed to be -0.52. This elasticity is applied to the price of crude oil rather than the price of gasoline, which would imply that it is too large and the response of quantity of gasoline demanded is therefore also too large. Taking these three factors together suggests that the CBER model is likely to overestimate the response of GHG emissions to the removal of highway fuel taxes by a large margin.

The third approach, NEMS-NAS, has the most detailed structure of any of the models (see the description in Appendix A). Additionally, and importantly, it has a detailed treatment of inter-fuel substitution and the different regulations and mandates, particularly the renewable fuel standards. It has the shortcomings of the other partial-equilibrium models of excluding spillover spending effects on outside the energy sector. Additionally, the price-elasticities are low compared to many studies.

The major and surprising result of the NEMS-NAS estimates is that the impact of removing the highway taxes is very small. Here is the reason for the surprising result. The key factor at work is the "volumetric bias" of highway fuels taxes. This signifies that the highway fuels taxes are imposed on a volumetric basis. Because ethanol, and particularly E85 (which contains 85% ethanol by volume), has a lower energy-to-volume ratio than gasoline, removing the highway-fuels excise taxes has the effect of favoring ethanol-based fuels. Additionally, because ethanol use is constrained by the renewable fuel standards (RFS), as described in detail in Chapter 5, the removal of highway taxes favors E85 over gasoline.

According to the NEMS-NAS simulations, total energy use in the transportation sector would rise by 0.32% over the period when highway fuels excise taxes are removed. However, because of the volumetric bias and RFS, gasoline use is slightly *lower* over the entire period, while E85 rises substantially. An increase in liquid fuel consumption would lead to an increase in CO_2 emissions except for the shift toward E85, which has a lower GHG emission rate than blended gasoline.

The committee notes an important reservation at this point concerning the increased use of E85 in these calculations as well as those in the FAPRI model below. E85 is used in Brazil, but it has not been in widespread use in the U.S. The NEMS-NAS calculations project a hundred-fold increase in the use of E85 over the next two decades. This increase is highly contingent on the RFS mandates continuing in force and on the development of an E85 car fleet and the associated fueling infrastructure.

The key result for NEMS-NAS is that removing the highway excise taxes results in a very small increase in CO_2 emissions of 3.5 MMT per year, or about 0.07% of average annual U.S. CO_2 emissions over the 2010-2035 period.

The FAPRI model has a detailed analysis of the structure of the biofuels industry and mandates. The petroleum sector is relatively aggregated. There are four separate markets in the petroleum sector: petroleum, gasoline, diesel, and residual oils. Petroleum is refined into the three petroleum products. Final demands consist of transportation, agriculture, and other. There is no detail of the transportation capital stock or fuel-efficiency standards. FAPRI has two regions, the U.S. and the rest of the world, and thus can calculate the impacts on global GHG emissions. Overall, FAPRI is well-designed to test for policies that work primarily through the biofuels subsidies and mandates, as well as the complex interactions of the different grades of ethanol. FAPRI assumes that the RFS mandates will apply (subject to waiving some of the advanced mandates). To meet the mandates requires that increasing amounts of E10 and eventually E85 will be produced and used in motor vehicles. As we note below, the growth rate of E85 use is extremely ambitious. Readers who wish to understand the full details should see Appendix A and the FAPRI model documentation referenced in Appendix A. For highway motor fuels tax, there was no sensitivity analysis performed on the effects of relaxing or removing the RFS mandates. Chapter 5 considers that sensitivity analysis for biofuels tax subsidies, and shows that the FAPRI and NEMS-NAS models have similar behavior when subsidies are removed.

FAPRI has a similar effect to NEMS because of the volumetric bias of highway fuels taxes and the RFS. The FAPRI model estimates that removing the highway fuels taxes would lead to an increase in GHG emissions of 9.8 MMT per year of CO_2-equivalent, or about 0.17% of U.S. CO_2 emissions over the 2014-2021 period. GHG emissions for the U.S. are slightly higher than the global total, while rest-of-the-world GHG emissions decline slightly when the highway taxes are removed. The basic factor leading to low GHG emissions is similar to that in the NEMS-NAS model: because ethanol has lower energy per gallon, reducing the highway excise taxes increases the use of ethanol relative to gasoline.

While the basic forces at work in FAPRI and NEMS-NAS are similar, FAPRI has a slight rise in gasoline consumption rather than the small decline in gasoline consumption in NEMS-NAS. The difference depends upon the time period and details of the specification and is probably not reliably resolved.

Table 4-2 summarizes these points on the four models in a succinct fashion.

AVIATION FUEL EXCISE TAX

As with highway fuels, the Internal Revenue Code imposes an excise tax on aviation fuels. Fuel for use in commercial aviation is taxed at $0.043 per gallon, while fuel for non-commercial use (that is, private use) is taxed at $0.193 per gallon for gasoline and $0.218 per gallon for jet fuel. In 2010 the IRS reported receipts of nearly $390 million from the tax on commercial use fuel and approximately $22 million from the tax on non-commercial fuel. For comparison, the IRC also imposes several taxes on passenger air transport with 2010 receipts totaling $7.6 billion.

Modeling Results

Only one of the modelers, CBER, estimated the impact of removing the tax on jet fuel. That model's projections suggest that cumulative CO_2 emissions would increase by over 70 MMT over the time period from 2010 to 2035, considerably less than their estimate of the impact of removing the highway fuels tax. On the other hand, the implied estimate of the change in government revenue per ton change in emissions is smaller for the jet fuel calculation than it was for the highway fuels excise tax, suggesting that for the same change in government revenue, a policy adjustment to air travel would have a bigger impact on emissions.

The CBER model assumes that the demand elasticity for jet fuel is the same as for all other transportation fuels: just above -0.5. The existing empirical literature on airlines suggests that the demand for air travel is more price elastic than the demand for gasoline and other oil products. Benchmark price-elasticity estimates for air travel are around -1 (Borenstein, 2011).

TABLE 4-2 Summary Appraisal of Studies of Impact of Removing Highway Fuels Taxes

Model	Modeling Period	Average CO_2 emmisions (MMT/year)	Advantages	Disadvantages	Net Appraisal
IGEM	2010-2035	83	General equilibrium approach; econometric estimates of many parameters	No gasoline or vehicle sector; no representation of regulations or biofuels mandates; no highway fuels taxes	Not applicable because of lack of sectoral details, highway fuels taxes, and mandates.
NEMS	2010-2035	3.5	Highly detailed structural model of vehicles and fuel sector; contains detail of RFS and ethanol products; vintage model of investment	Partial equilibrium; U.S. emissions only.	Most appropriate modeling approach. Result depends critically on the presence of volumetric bias and renewable fuel standards.
CBER	2010-2035	54	Transparent; some interfuel substitution; world petroleum market	Long-run elasticities; no reprentation of biofuels mandates; elasticities high; partial equilibrium	Likely to overestimates he impact on GHG emissions by a large factor.
FAPRI	2014-2021	9.8	Highly detailed model of biofuels sector; global impacts.	Highly stylized treatment of petroleum demand; partial equilibrium.	Appropriate modeling approach. Result depends critically on the presence of volumetric bias and renewable fuel standards.

How a change in air travel will translate into a change in the amount of jet fuel consumed is complicated, and depends on the length of flights, number of takeoffs and landings, aircraft used and other factors. Also, changes to the price of jet fuel could eventually cause airlines to make adjustments to the way they do business, including changing the speed of flights, the number of flights, capital investments in existing aircraft (such as the installation of winglets) and the fuel efficiency of the aircraft used. Understanding the relationship between fuel prices and the fuel efficiency of air travel remains an important area for future research. If one were to account for more elastic demand for air travel as well as adjustments made by airlines in response to changes in their input costs, the implied impact of changes to jet fuel taxes would likely be larger than reflected in the CBER modeling runs.

Summary on Aviation Taxes

While the total GHG impact of removing the tax on jet fuel is small, jet fuel taxes are a small component of the total taxes on air travel. For example, the federal ticket tax is 7.5 percent of federal revenue from excises, while fuel taxes are less than 0.5 percent of revenue. Adjustments that reflect taxes in addition to the tax on jet fuel would have a commensurately larger impact on GHG emissions, although they may not have the same impact on airlines' decisions about fuel efficiency. Given the potential impacts in this sector, it remains an important area for future research.

OVERALL EVALUATION ENERGY-SECTOR EXCISE TAXES

This chapter has reviewed research on the impacts of excise taxes in the energy sector on greenhouse-gas emissions. There are two important sets of excise taxes—those on highway fuels and those on air travel. Most of the research in this and earlier studies has focused on the taxation of highway fuels, and this summary pertains primarily to that sector.

This chapter reviewed four commissioned studies of the effect of removing the excise taxes on highway fuels. (For a discussion of the different models and their treatment of fuel demands, see Appendix A.) All four models find that removing the excise taxes on highway fuels would result in increasing greenhouse gas emissions. This result occurs because a lower post-tax price for highway fuels generates higher demand for highway fuels, which are largely derived from petroleum products.

But the magnitude of the estimated effects varies dramatically for the different models. The committee notes that large differences in projections of different energy-economic models have been seen in other model-comparison studies, so there is ample precedent for divergent results.[4] Having studied the model results and the broader literature, the committee concludes that the differences among the models are large and incompletely understood. The differences arise from the types and values of price elasticities used by the different models, from assumptions about increasing biofuels production and consumption to meet the RFS mandates, from the volumetric bias of highway fuels taxes, and from application of the tax within each model's structure. A close examination of the results leads the committee to conclude that the NEMS-NAS and the FAPRI models capture the forces at work in this sector most reliably and therefore form the basis of our estimates. Taking these two modeling results together produces a striking conclusion: The impact of removing highway fuels taxes on GHG emissions is estimated to be very small because of special features of the taxes and the market. The volumetric bias of the taxes means that removing the taxes favors ethanol, which will reduce the GHG impacts of increasing highway fuel consumption. Additionally, the renewable fuel standards constrain the use of ethanol. According to the two models, the effect of removing the highway fuels taxes is 4 MMT per year (NEMS) and 10 MMT per year (FAPRI). These are 0.07% and 0.17% of annual U.S. CO_2 emissions, respectively. The third model (CBER) is similar to these two models when adjustments are made to account for the upward bias in its methods.

The committee emphasizes the contingent nature of the model projections. They are contingent because the results depend upon the structure, timing, and implementation of the renewable fuels standards (RFS) as well as a quirk in the tax structure (its volumetric bias). Moreover, the impact works through E85, which has not yet entered significantly in U.S. fuel consumption. If the RFS were to disappear tomorrow, or if the regulations on E85 were to change drastically, or the highway motor fuels taxes were levied by energy content instead of by volume, the projected impacts of removing the gasoline tax might be substantially different and would probably be significantly larger. Additionally, as discussed in the next chapter, there are many uncertainties about how the most recent version of the RFS (RFS2) will be implemented. Finally, it should also be noted that the complex structure of the RFS may imply that large tax increases will have different effects from the small tax decreases that the current study examined.

The magnitude of the differences between models leads the committee to caution against relying on specific numerical results from a single model and recommends drawing only broad conclusions about the nature and direction of impacts. Policy makers and analysts should rely on multiple models, methodologies, and estimates in calculating impact of the tax code and other policies on greenhouse-gas emissions and climate change.

---

[4]The Energy Modeling Forum (EMF) at Stanford University has undertaken several model comparison studies in energy, oil, and climate change. Projections of energy consumption in the recent EMF-22 comparison were found to differ by almost a factor of two among models between 2010 and 2050 (see Clarke et al. 2010).

Chapter 5

Biofuels Subsidies

INTRODUCTION

This chapter examines the greenhouse gas (GHG) impacts of federal biofuels policies: excise tax credits for ethanol and biodiesel, the tariff on ethanol, and the federal renewable fuels standard. Although the tax credits and the tariff have expired, the renewable fuels standards are playing an increasingly important role in the motor fuels sector. The committee devoted considerable attention to the taxation and regulation of biofuels for three principal reasons. First, ethanol credits have been widely used in the United States and abroad, have been among the largest energy-related tax expenditures in revenue foregone, and their impacts on GHG emissions are important public policy questions. Second, the results illustrate the often-unintended impact of tax expenditures, because of the complexity of the regulatory and interindustry feedbacks. Third, the biofuels standards interact significantly with motor fuels taxes and the use of petroleum.

The tax provisions need to be analyzed in the context of the regulatory framework. The federal mandates for biofuels production arising from the Energy Policy Act (EPAct) of 2005 and the Energy Independence and Security Act (EISA) of 2007 established requirements for the volume of renewable fuels that must be blended into transportation fuels. EISA, the currently binding policy, schedules the amount of required biofuels to increase from 9 billion gallons in 2008 to 36 billion gallons by 2022. The committee's analysis finds that these regulatory mandates severely constrain the magnitude of the impacts of the tax incentives.

FINDINGS FROM PRIOR LITERATURE

Most studies analyzing the impacts of biofuels do not directly consider the GHG effects of specific tax code provisions. There are, however, several studies that consider important interactions between the tax code, renewable fuels mandates, and crop price supports (Gardner, 2007, and Schmitz, 2007). Those studies find that the Renewable Fuel Standard (RFS) mandates are more effective than the tax incentives, and furthermore that the RFS effectively limited the impact of the tax incentives on renewable fuels production and consumption (de Gorter, 2008). One study also found that the crop price supports for ethanol feedstocks, such as corn, combined with quantity mandates for ethanol may lead to an increase in petroleum consumption, similar to the results of our modeling efforts reported below (de Gorter, 2010).

Beyond these studies, much of the rest of the literature focuses on questions of whether or not ethanol production and consumption leads to a net increase or decrease of GHGs per Btu of fuel (75 Fed. Reg. 14760 [2010]; Yacobucci, 2010; Gelfand, 2011). While not directly linked to the impacts of specific tax provisions, such literature is still informative in determining whether those impacts are likely to be net positive or net negative (Mosnier et al., 2013).

PROVISION-BY-PROVISION ANALYSIS BIOFUELS CREDITS AND ETHANOL TARIFF

Legal Description

Prior to 2013 the Internal Revenue Code (IRC) provided three income tax credits for alcohols used as a motor fuel. Fuel alcohols blended with gasoline or used pure as a fuel both qualified for a $0.45 per gallon credit under the Volumetric Ethanol Excise Tax Credit (VEETC). Gasoline suppliers that blend ethanol into their fuel could take this credit as an instant rebate against motor fuels excise tax liability or as a nonrefundable credit against their income tax liability, if any, for a given year. In practice, nearly all taxpayers preferred to claim the excise credit. Doing so immediately captured the credit's benefits and eliminated the risk of not having sufficient income tax liability to fully utilize the credit.

While the law made the credit available to several types of alcohol, ethanol was and remains the principal alcohol used as motor fuel in the United States. Ethanol from small producers qualified for a $0.10 per gallon credit. This credit was limited to the first 15 million gallons of annual ethanol production from producers capable of distilling less than 60 million gallons per annum. Producers of cellulosic biofuels received a $1.01 per gallon income tax credit until December 31, 2012, one year after the expiration of the general ethanol income tax credits. The cellulosic producer credit is commensurately reduced by the amount of any other tax credits applied to the fuel. For instance, if the VEETC is applied to the blender, the net producer credit is $1.01 − $0.45 = $0.56 per gallon. Cellulosic biofuels are defined as any liquid fuel produced from any lignocellulosic or hemicellulosic matter that is available on a renewable or recurring basis. Common sources for cellulosic biofuels include switchgrass, corn stover, and wood chips.[1]

The IRC provided similarly structured credits for biodiesel until December 31, 2011. Each gallon of biodiesel was eligible for a $1.00 per gallon credit while small agri-biodiesel producers, defined by the same volumetric limits as small ethanol producers, were eligible for a $0.10 per gallon credit. Agri-biodiesel refers to biodiesel made using virgin oil instead of reclaimed oil.

Ethanol Tariff

In addition to tax credits, a $0.54 per gallon ethanol tariff on imported ethanol historically benefitted the U.S. ethanol industry by reducing the competitiveness of imported ethanol. The tariff was originally intended to prevent imported ethanol from benefitting from the U.S. tax credit. The ethanol tariff expired on January 1, 2012. The tariff's expiration will primarily benefit Brazil, which has a large ethanol industry based on sugarcane. The energy produced by ethanol compared with the energy invested in its production (the energy return on energy invested) is higher for the sugarcane ethanol produced in Brazil than it is for conventional (corn-based) ethanol in the United States (75 Fed. Reg. 14760 [2010]). Thus, from a global perspective, the expiration of the ethanol tariff can be expected to increase the Brazilian sugarcane share of the U.S. ethanol market and thereby decrease the GHG emissions from ethanol fuels used in the United States.

Pathway to GHG Impact

The various tax credits lowered the cost of biofuels and therefore should have encouraged their substitution for petroleum motor fuels. Because biofuels are almost always sold as a blend with petroleum fuels, however, the subsidies also effectively lowered the final delivered price of the petroleum-biofuel blend, thereby encouraging additional consumption of petroleum. Although the literature shows a range of estimates for life-cycle GHG emissions from all biofuels, depending on whether agricultural practices and soil-based carbon is considered, most studies suggest reduced emissions for biofuels compared with petroleum-based analogs. Cellulosic ethanol shows significantly greater GHG reductions than corn-based ethanol, and may not be subject to the food-fuel substitution criticism often leveled at corn-based ethanol. The Environmental Protection Agency (EPA) estimates a much greater reduction in life-cycle GHG emissions from cellulosic ethanol than the reduction in life-cycle GHG emissions from

---

[1]Kelsi Bracmort, Randy Schnepf, Megan Stubbs, and Brent D. Yacobucci, Cellulosic Biofuels: An Analysis for Congress, Cong. Res. Serv. Rep. RL34738 (Oct. 14, 2010), at 1. Internal Revenue Code Sections 40, 40A, 4041, 4081, 6426 and 6427(e).

corn-based ethanol. The analysis in this chapter uses the standard EPA emission factors for different fuel types (75 Fed. Reg. 14760 [2010]) to estimate the net GHG effects and examines the sensitivity of the results to variation of these emission factors.

Fiscal Impact

Expenditures on biofuels subsidies represent only a small tax expenditure if measured solely by lost income tax revenue. Ignoring the recently expired excise tax credits, the Treasury Department and Joint Committee on Taxation estimate the 2010 tax expenditures on ethanol and biodiesel at $90 million and $100 million, respectively. Including the subsidy provided through the excise tax system, those estimates increase to $6.26 billion and $5.2 billion. Either of these estimates was the largest of the energy-related tax expenditures estimated by either group, and though small compared with broad-based tax expenditures such as that for the exclusion of employer-sponsored health care, are still sizable impacts to the Treasury.

Renewable Fuels Standard

Although not a tax provision, the Renewable Fuels Standard is an important set of regulatory mandates that substantially affect biofuel use in the United States and thus must be included in the evaluation of the tax provisions outlined above. The RFS, created by Congress under the Energy Policy Act of 2005 (P. L. 109-58), established the nation's first mandate for renewable liquid fuels. The original RFS program (commonly referred to as RFS1) required 7.5 billion gallons of renewable fuel to be blended into gasoline by 2012. The mandate was expanded by the Energy Independence and Security Act of 2007 (P.L. 110-140), and is now commonly referred to as RFS2. RFS2 expansion of the program included the following:[2]

- It included diesel, in addition to gasoline;
- It increased the volume of renewable fuel required to be blended into transportation fuel from 9 billion gallons in 2008 to 36 billion gallons by 2022;
- It established distinct categories of renewable fuel, and sets separate annual volume requirements for each one (by 2022):
  o Conventional biofuels (e.g., corn-based ethanol): 15 billion gallons maximum

---

[2]More information on the program can be found in the Congressional Research Service Report (Schnepf and Yacobucci, 2013) or on the EPA Web site (http://www.epa.gov/otaq/fuels/renewablefuels/index.htm).

    o Advanced biofuels: 21 billion gallons minimum, including the fol-
lowing minimums for specific advanced categories:
- Cellulosic ethanol (16 billion gallons)
- Biodiesel (1 billion gallons, or to be determined by U.S. EPA)
- Other advanced biofuels (e.g., sugarcane ethanol) may fill the gap
between the cellulosic and biodiesel minimums and the 21 billion
gallon total advanced biofuel minimum.
- It required EPA to apply GHG performance thresholds for each renewable fuel category, so that each fuel would have demonstrated
lower GHG emissions than the petroleum fuel it replaces (e.g., gasoline or conventional diesel).

This study examines the interaction of the RFS with the tax provisions by
initially assuming that the RFS2 will remain in effect as stipulated by current
law. Then alternative modeling results will be generated assuming the RFS is
not in place in order to gauge the interaction effects between the tax provisions
and the RFS mandates.

Modeling Approach and Key Assumptions

A large variety of approaches have been used to examine the economics of
biofuels and biofuel policy. These include general equilibrium models (Gurgel et
al., 2007; Tyner et al., 2010; Decreux and Valin, 2007), agricultural optimization
models (Adams et al., 1996; Beach and McCarl, 2010), simulation models (Wise
et al., 2009), and econometric-based simulation models (Babcock and Carriquiry,
2010).

There are several challenging aspects of modeling biofuel policy: (1) the
complex interactions with agriculture and agricultural policy, including competing demands for crops and by-products supplies of animal feeds; (2) the complex policy requirements of the Renewable Fuel Standard (RFS2, as described
below) and their interaction with investment and production tax credits that differentially treat different biofuel production pathways and feedstocks that are the
focus of this report; (3) international linkages in agriculture and energy markets;
(4) land-use change and competition for land; and (5) the carbon implications of
land-use change.

This chapter reports results of variations in biofuel-related federal tax provisions simulated using two different models: (1) the Food and Agricultural Policy Research Institute at the University of Missouri (FAPRI-MU) model and (2)
the National Energy Modeling System as run by OnLocation, Inc., for the National Academy of Sciences (NEMS-NAS). The chapter will focus initially on
FAPRI-MU, because of its unique coverage and detail of the U.S. and world
agricultural and motor fuel sectors. NEMS-NAS results will be shown as a sensitivity case later in the chapter.

The FAPRI-MU model as employed here is a system of demand and supply functions for 16 crops, 15 crop-based products, and 17 different types of livestock and livestock-based products (Meyers et al., 2010; Devadoss et al., 1993). Some of these functions are econometrically (statistically) estimated from historical data, while other functions are based on assumed forms and parameter values. In particular, the rapid changes in biofuel markets make direct estimation based on observed behavior difficult. A good example is E85 demand, which can be very important in future market projections but has not accounted for more than a very small amount of biofuels consumption in the past. The estimates are updated periodically using the most recent available data.

FAPRI-MU's focus is on the United States, with the rest of world either collapsed into a single rest-of-world supply-and-demand response, as in the case of animal products; composed of a similar rest-of-world aggregate response but with key countries identified, as in the case of ethanol; or represented with aggregate rest-of-world supply-and-demand aggregates, as in the case of main crops. This longstanding agricultural model has been recently augmented to include detailed modules on oil markets (Thompson et al., 2011) and U.S. biofuels markets (Thompson et al., 2008). The strength of the model is in its detailed representation of agricultural markets, including global markets, modeling of the complex Renewable Identification Number (RIN) fuel credits with multiple fuel production pathways representing both conventional and advanced-generation processes, and links to global petroleum and refined fuel markets (Thompson et al., 2010).

The FAPRI-MU approach does not explicitly consider land use or the carbon implications of land-use change. These are highly uncertain responses with wide-ranging results in the literature (Plevin et al., 2010; Searchinger et al., 2008; Keeney and Hertel, 2009; Tyner et al., 2010; Hertel, 2011; Mosnier et al., 2013; and Melillo et al., 2009). Instead, the greenhouse gas implications of alternative policies are assessed by applying a fixed GHG coefficient per unit of fuel for different biofuel production pathways. Calculations are based on three sets of GHG emission factors based on differing assumptions about life-cycle energy use and indirect land-use change factors (ILUC). The default coefficients are the thresholds values stipulated in the EISA legislation for each fuel type. Alternative coefficients are evaluated in a sensitivity analysis included in the results discussion below.

The FAPRI-MU model is benchmarked to government projections for energy and agriculture, and is resolved annually for a 10-year period (2011 to 2021). For energy, the model is benchmarked against the Energy Information Administration's (EIA) 2012 Annual Energy Outlook (AEO) for petroleum and refined oil markets until 2011, but AEO 2011 petroleum prices and other key variables guide the projections for consistency with the other modeling analyses undertaken for this study (U.S. EIA, 2012). EIA's Outlook assumes gross domestic product (GDP) grows at an annual average rate of about 2.6 percent through 2022, crude oil prices rise from $109 per barrel in 2013 to $135 per barrel in 2021, and gasoline prices rise from about $3.40 to $4.40 per gallon. On the agricultural side, the short-run projections are calibrated to the U.S. Department of Agriculture's World Agricultural Supply and Demand Estimates. Projections of market components outside of those two sources are calibrated to an early version of the FAPRI-MU agricultural baseline (Westhoff et al., 2012), but the model was resolved with an extension of U.S. biofuel blenders credits, ethanol-specific duty, and cellulosic producer credit to generate the baseline used here. The corn price is a key input to ethanol and illustrative of crop prices in general. It rises from $5.16 to 5.50 per bushel over the analysis period of 2014 to 2021.

To be consistent with other chapters in the report, Table 5-1 outlines assumed values used in the modeling analysis for several key factors.

ANALYSIS OF VOLUMETRIC ETHANOL EXCISE TAX CREDIT (VEETC), BIODIESEL BLENDER CREDIT, CELLULOSIC BIOFUEL PRODUCER CREDIT, ETHANOL-SPECIFIC DUTY

Modeling Results

The FAPRI-MU model, described above, is used to estimate the impacts of the identified biofuel provisions on GHG emissions and other key variables. The model simulates for the period 2011–2021 key outcomes for three core scenarios:

1. **Reference Scenario:** Continuation of policies (RFS2 and other) as they were in effect at the time of analysis (March 2012) and expected changes in energy and agricultural technology, markets, and macroeconomic factors.
2. **Remove VEETC:** Identical to *Reference* except that the Volumetric Ethanol Excise Tax Credit is eliminated.

3. **Remove all Provisions**: Identical to *Remove VEETC* except the Biodiesel Blender Credit, Cellulosic Biofuel Producer Credit, and Ethanol-specific Import Duty are also eliminated.

TABLE 5-1 Key Modeling Assumptions

Assumptions	
Modeling Period	2014-2021
Real GDP Growth (% per year)	2.6%
Oil Price (2014-2021, avg nominal)	$123.47
Energy Consumption Growth	From 2012 AEO Outlook
GHG Included	CO_2, N_2O, CH_4
U.S. or World?	World and U.S.

In this chapter, as elsewhere, we report the results of the calculations with one- or two-digit precision because that is how the numbers are reported by the economic models. The committee notes that the numerical precision of the calculations does not imply a corresponding accuracy of the projections or estimates. As we note elsewhere, the committee did not prepare statistical uncertainty analyses of the estimates. Comparisons across models indicate that the uncertainties are large. So readers should be alerted that reporting model output at a specific precision does not imply that the actual results are similarly precise.

The reference scenario represents a projection that assumes a particular set of policies, technologies, economic, and demographic phenomena during the projection period of the model (2011-2021) and provides the starting point for analysis of alternative policies.

Of particular note, the current renewable fuels standard (RFS2) is assumed to remain in place during the entire simulation period. However, the modeling exercise assumes that the cellulosic ethanol component of the mandate will be waived each year, as it has since inception of the program, due to insufficient production capacity. The exercise assumes the EPA resets the cellulosic waiver amount to the level of output that would be produced economically in response to the market price for cellulosic ethanol, the separate renewable fuels credit (RIN) price, and the value of the applicable tax credits. This will generally cause production to drop below the mandated level. A share of the cellulosic shortfall (25 percent of it) is assumed to be met by other advanced biofuels in the future, and the remainder reduces the total mandate accordingly.

The first reference scenario, *Remove VEETC,* focuses on removing the largest of the biofuel tax provisions, VEETC, which provides fuel blenders a per-gallon tax credit for using ethanol. The second reference scenario, *Remove all Provisions,* assesses the elimination of all three major biofuel tax code provisions, along with the ethanol-specific import duty. The duty, although not explicitly part of the tax code, was included because of its potential effect on federal revenues and on the composition of biofuels used and the GHG consequences thereof.

Table 5-2 provides a summary of key modeling results for the baseline and biofuel policy scenarios. The results are summarized by category.

GHG Effects

Removing VEETC alone is projected to lead to a roughly 5 MMT reduction in GHG emissions, globally. There are three noteworthy aspects to this estimate.

First, 5 million tons is a very small number, roughly 0.1 percent of total U.S. GHG emissions (U.S. Environmental Protection Agency, 2012), or about one-fifth of the emissions of one very large coal-fired power plant in the United States (Center for Global Development, 2007). Thus, the VEETC provision does not appear to have a meaningful impact on GHG emissions.

Second, the estimate is global, as it takes into account emissions in other parts of the world due to (1) fuel market feedback effects, wherein subsidized ethanol lowers blended fuel prices, thereby increasing the demand for gasoline, which increases consumption and emissions on the margin; and (2) indirect land-use change effects, in which subsidized ethanol leads to more agricultural feedstock used for fuel, which diverts it from other uses such as food, which leads to agricultural intensification and land clearing, which generates emissions. The fuel market feedback effect is captured in Thompson et al., 2011. The ILUC effects, as discussed in the chapter introduction are determined by exogenous emissions coefficients in FAPRI-MU and can vary widely. As a result, a sensitivity analysis to these coefficient values is presented further below in this section.

The third and most striking result is that the calculations indicate that the VEETC actually *increases* GHG emissions. The sensitivity analysis using alternative GHG emission coefficients below will show that this result is sensitive to alternative assumptions and that the impact of the VEETC is generally very small regardless of its sign (the first point above).

The second policy scenario that removes all biofuel provisions has a very small incremental impact on GHG emissions, with a central estimate of an additional 0.6 MMT emissions reduction if the three other biofuel provisions were also removed. One reason for this is that the RFS2 standard is still in place when the tax provisions are removed. Thus, total biofuel use is only marginally affected by the removal of the provisions, though, as shown below, the mix of biofuels to meet the target can change in response to the provisions being dropped.

Revenue Effects

Removing VEETC would lower federal tax expenditures by approximately $7.2 billion per year between 2014 and 2021 (were the former policy to be in place during this period). This savings comes in the form of reduced tax credits issued to fuel blenders for their use of ethanol. Removing all provisions would reduce expenditures from the Treasury by about $12.6 billion per year, as payouts for biodiesel and cellulosic production are eliminated. This saving is reduced slightly by the reduction in tariff receipts from imported ethanol as that provision is dropped.

Table 5-2 includes a calculation of the tons of emissions generated (or avoided) per dollar of federal revenue affected. The values are 0.0007 and 0.0004 for the *Remove VEETC* and *Remove all Provisions*, respectively. The fact that these numbers are positive reflects that every dollar of federal revenue (that is, foregone in tax receipts) generates a small *increase* in emissions. One might have expected that a policy intending to reduce GHG emissions would,

TABLE 5-2 Removal of Biofuel Provisions – Key Modeling Results

Key Variable (annual average, 2014-2021)	Baseline (with RFS2)	Remove VEETC Change Relative to Reference Scenario	%	Remove all Provisions Change Relative to Reference Scenario	%
CO_2-e Emissions (MMT)		-4.8		-5.4	
Federal Expenditures ($ billion)		-7.2		-12.6	
Tons CO_2-e per $ of Revenue (calculated)		0.0007		0.0004	
FUEL USE (billion gallons, gasoline equivalent)	**Change Relative to Reference Scenario**	**Change Relative to Reference Scenario**	**%**	**Change Relative to Reference Scenario**	**%**
Gasoline Use					
World	360.10	+0.44	+0.1%	+0.64	+0.2%
U.S.	120.91	+1.29	+1.1%	+1.30	+1.1%
Ethanol Use					
World	27.75	-1.15	-4.1%	-2.35	-8.5%
U.S.	15.27	-1.60	-10.5%	-1.72	-11.3%
Conventional	11.90	-1.74	-14.6%	-1.75	-14.7%
Cellulosic	2.33	0.69	+29.5%	-1.58	-67.9%
Other Advanced	1.04	-0.55	-52.7%	+1.61	+154.4%
Diesel Use					
World	445.14	+0.25	+0.1%	+0.42	+0.1%
U.S.	73.20	-0.08	-0.1%	-0.11	-0.2%
Biodiesel Use					

World	9.43	0.00	—	-0.01	-0.1%
U.S.	1.31	0.00	—	0.00	—
U.S. FUEL PRICES (wholesale $ per gallon unless otherwise indicated)					
Petroleum Refiner's Cost ($/barrel)	123.47	+0.63	+0.5%	+0.91	+0.7%
Gasoline	3.52	+0.06	+1.7%	+0.07	+2.0%
Ethanol					
Conventional	2.82	-0.30	-10.5%	-0.26	-9.1%
Cellulosic	4.12	+0.21	+5.0%	-0.63	-15.3%
Other Advanced	3.27	-0.24	-7.4%	-0.07	-2.2%
Diesel	2.90	+0.02	+0.7%	+0.03	+1.0%
Biodiesel	5.51	-0.09	-1.6%	-0.11	-2.0%
CROP AREA (MM ac)					
World					
Corn	439.9	-4.27	-1.0%	-3.26	-0.7%
Soybean	278.9	+1.27	+0.5%	+1.09	+0.4%
U.S.					
Corn	93.8	-3.32	-3.5%	-2.47	-2.6%
Soybean	73.4	+1.15	+1.6%	+1.05	+1.4%
U.S CROP PRICES ($/bushel)					
Corn	5.42	-0.34	-6.3%	-0.30	-5.5%
Soybeans	11.98	-0.23	-2.0%	-0.27	-2.2%

instead, generate a negative value for tons per revenue, giving a sense of the tradeoff between federal expenditures and emissions outcomes—the more spent, the lower the emissions. For the best-guess estimates of the parameters used in this study, the provisions lead to both revenue losses and higher GHG emissions.

Fuel Consumption Effects

As discussed below, removing the biofuel provisions changes the relative prices of motor fuels. This causes substitution among fuel types as indicated in Table 5-2. All fuel substitution reflects differences in energy content of the different fuels (e.g., gasoline has higher energy content than ethanol). Ethanol use, of course, declines with the removal of the VEETC and gasoline use increases. The ethanol use changes are fairly substantial—world use declines 4.1 percent for all ethanol types, while U.S. use drops 10.5 percent. In the United States, the largest absolute reduction is in conventional (corn-based) ethanol, which is reduced by 1.7 billion gallons per year (about 15percent).

There is a projected increase in cellulosic ethanol of 690 million gallons per year (30 percent), as the RFS mandate remains in effect and the cellulosic ethanol subsidy remains in place and is increased automatically under current law if the VEETC is eliminated. That is because, as indicated above, the cellulosic producer's credit ($1.01 per gallon) decreases by the amount of an existing VEETC ($0.45). When the VEETC is eliminated, the cellulosic credit returns to the full $1.01 value. Since both the cellulosic subsidy and import tariff (mostly on advanced sugarcane biofuel imports from Brazil) remain in effect, much of the increase in cellulosic is countered by a decline in advanced biofuels of 550 million gallons (53 percent). The substitution effects are strengthened by the need to meet the tiered RFS2 mandate.

Changes in the use of diesel—conventional and biodiesel—are quite small in proportion: less than a 0.2 percent increase.

The scenario removing all the biofuel provisions affects the scale and distribution of effects. The overall absolute effects tend to be larger when all provisions are eliminated as more of the fuel base is affected. However, cellulosic ethanol declines substantially (68 percent) when all of the provisions—including the cellulosic subsidy—are dropped, in contrast to the large rise when VEETC alone is removed, as discussed above. Alternatively, advanced biofuels rise in use more than 150 percent when all the provisions are dropped, primarily because one of those provisions, the ethanol import tariff, is essentially a barrier for imported advanced biofuels from Brazil, and because of the assumption about the implementation of a cellulosic mandate waiver. This creates an almost complete substitution of advanced imports for domestic cellulosic ethanol to meet the RFS2.

Fuel Price Effects

Removing the biofuel provisions raises the price of petroleum and gasoline slightly, because the demand for those products rises when the biofuel tax preferences disappear and ethanol production declines. The effects are larger when all provisions are removed, but all effects are no more than 2 percent of total production. At the same time, the price of biofuels declines as blenders lose tax preferences and, correspondingly, willingness to pay for biofuels. An exception is that the price of cellulosic ethanol is projected to rise 5 percent when VEETC only is removed, because of the induced shift in demand discussed above. Further, the advanced ethanol price declines in both scenarios, but it declines less in the *Remove all Provisions* scenario, because the shift in demand for advanced ethanol caused by removal of the import tariff provides some price pressure on those imports, though not enough to counter the decline in price caused by removing the tariff in the first place.

Crop Market Effects

The biofuel provisions generally raise the demand for corn as the dominant feedstock for conventional ethanol. They do so in part by expanding crop area and by inducing growers to shift land from other crops, principally soybeans to corn. As a result, removing the provisions would lead to a decline in corn acreage and an increase in soybean acreage. Globally these effects are small, and less than 1 percent of the crop base is affected. Within the United States, the effects are larger, with a decline of between 2.5 and 3.5 percent in corn acreage and an increase of 1.5 percent in soybean acreage.

Correspondingly, corn prices are expected to decline as the demand for corn drops. Soybean prices drop as well, as land shifts back into soybean production and soybean production increases. The projected decline in corn prices is about 5.5 to 6 percent and the decline in soybean prices is 2 to 2.5 percent. These effects are not insubstantial and, while not the focus of this study, they could have a substantial (positive) impact on household budgets through a decrease in food-related costs (Chakravorty et al., 2012; Hertel et al., 2010; and Roberts and Schlenker, 2010).

Key Interaction Effect: Renewable Fuels Standard

The modeling results above all assume that the RFS2 mandate remains in place. We now examine the importance of that assumption by testing the results against a scenario in which the renewable fuel standards are removed. That is, we analyze the results of the *Remove all Biofuels Provisions* scenario with and without the RFS2 mandate in place. This comparison of results is found in Table 5-3. Note that Table 5-3 has two separate columns for reference scenario values, as the when the RFS2 is assumed to hold for the projection period.

We find that removing the RFS2 mandates amplifies the effects of removing the biofuel provisions. In essence, the full substitution away from biofuels to conventional fuels when tax preferences are lifted is more limited when the RFS2 is in place, because the mandates must be met by the biofuels even when their cost advantages are reduced. There are a few exceptions to this rule. The effects of provision removal is smaller in the cases of advanced and cellulosic ethanol when there are no RFS2 mandates, primarily because there is very weak demand for those products when the mandate is removed and thus much less substitution among ethanol types.

SENSITIVITY ANALYSES

Sensitivity Analysis 1: Biofuel Net GHG Coefficients

The net GHG effects of each biofuel component are exogenously determined via coefficients used in the FAPRI-MU model. As discussed above, the central coefficients used to generate the results presented above were derived from the per-fuel thresholds established in the EISA legislation. Due to inherent uncertainty about key factors such as indirect land-use change, the model was rerun with different coefficients to gauge sensitivity to this and other factors, including considering the source of the estimate, EPA v. EISA, and alternative estimates of the biofuel emission factors based on the literature. The different coefficients capturing life-cycle emissions per gallon of gasoline are listed in Table 5-4.

Results of the GHG coefficient sensitivity analysis are presented in Table 5-5 for the *Remove all Provisions* scenario. Note that the EISA column matches the net GHG results in Table 5-3. Using the EPA estimates, emissions can either decline slightly or increase slightly depending on whether ILUC emissions are included. Looking at the range of estimates in the literature, the "high biofuel emissions" estimates incorporate higher ILUC values and thereby more pronounced emission reduction benefits if biofuel provisions are removed. The opposite is the case when "low biofuel emissions" are assumed, in which case dropping the biofuel provisions would lead to a net increase in GHG emissions. The high-low range is wide in relative terms, but small in absolute terms.

Sensitivity Analysis 2: NEMS-NAS Model

To examine how differences in modeling approach might change the nature of findings, the committee also used the NEMS-NAS model (see Chapter 3) to analyze impacts of the biofuel provisions. Unlike FAPRI-MU, NEMS-NAS does not capture international market effects for motor fuels or agricultural feedstocks, nor does it estimate the effects of ILUC. However, NEMS-NAS does

TABLE 5-3 Effect of the RFS2 Mandate on Model Projections for the "Removing All Biofuel Provisions" Scenario: Key Modeling Results

Key Variable (Annual Average, 2014-2021)	With RFS2 Mandates			No RFS2 Mandates		
	Reference Scenario	Change Relative to Reference Scenario	%	Reference Scenario	Change Relative to Reference Scenario	%
CO₂-e Emissions (MMT)		-5.4			-7.0	
Federal Expenditures ($ billion)		-12.6			-10.1	
Tons CO₂-e per $ of Revenue		0.0004			0.0007	
FUEL USE (billion gallons, gasoline equivalent)						
Gasoline Use						
World	360.10	+0.64	+0.2%	360.55	+0.76	+0.2%
U.S.	120.10	+1.30	+1.1%	122.56	+1.75	+1.4%
Ethanol Use						
World	27.75	-2.35	-8.5%	26.24	-2.40	-9.2%
U.S.	15.27	-1.72	-11.3%	13.50	-2.15	-16.0%
Conventional	11.90	-1.75	-14.7%	12.51	-1.85	-14.8%
Cellulosic	2.33	-1.58	-67.9%	0.64	-0.63	-97.9%
Other Advanced	1.04	+1.61	+154.4%	0.35	+0.32	+93.3%
Diesel Use						
World	445.14	+0.42	+0.1%	445.3	+0.47	+0.1%
U.S.	73.20	-0.11	-0.2%	71.97	-0.27	-0.4%
Biodiesel Use						

(Continued on page 106)

World	9.43	-0.01	-0.1%	8.49	-0.15	-1.7%
U.S.	1.31	0.00	—	0.36	-0.14	-39.0%
U.S. FUEL PRICES (wholesale $ per gallon unless otherwise indicated)						
Petroleum Refiner's Cost ($/barrel)	123.47	+0.91	+0.7%	123.89	+1.12	+0.9%
Gasoline	3.52	+0.07	+2.0%	3.59	+0.10	+2.7%
Ethanol						
Conventional	2.82	-0.26	-9.1%	2.88	-0.28	-9.8%
Cellulosic	4.12	-0.63	-15.3%	3.44	-0.84	-24.5%
Other Advanced	3.27	-0.07	-2.2%	2.88	-0.28	-9.8%
Diesel	2.90	+0.03	+1.0%	2.89	+0.03	+1.2%
Biodiesel	5.51	-0.11	-2.0%	3.67	-0.87	-23.6%
CROP AREA (MM ac)						
World						
Corn	439.9	-3.26	-0.7%	441.70	-4.53	-1.0%
Soybean	278.9	+1.09	+0.4%	276.56	+1.25	+0.5%
U.S.						
Corn	93.8	-2.47	-2.6%	95.22	-3.45	-3.6%
Soybean	73.4	+1.05	+1.4%	71.78	+1.22	+1.7%
U.S. CROP PRICES ($/bushel)						
Corn	5.42	-0.30	-5.5%	5.45	-0.39	-7.2%
Soybeans	11.98	-0.27	-2.2%	11.47	-0.38	-3.4%

TABLE 5-4 Alternative Emission Coefficients for Gasoline and Biofuels Based on Study

Emission Coefficients	EISA	EPA, with ILUC	EPA, No ILUC	High Biofuel Emissions	Low Biofuel Emissions
		(kg CO$_2$-eq/gallon gasoline equiv)			
Gasoline	12.3	12.3	12.3	12.3	12.3
Diesel	12.2	12.2	12.2	12.2	12.2
Ethanol					
Conventional	9.9	9.9	6.2	12.3	6.2
Cellulosic	4.9	2.1	2.0	8.5	0.0
Other Advanced	6.2	4.8	4.1	6.2	4.1
Biodiesel	6.1	5.3	1.0	6.1	1.0

Key: EISA – Energy Independence and Security Act threshold values for biofuels; EPA – EPA Regulatory Impact Analysis of Renewable Fuels Standard; ILUC – Emissions from indirect land-use change; and High (low) biofuel emissions – expert judgment of high (low) emission values, based on literature.

TABLE 5-5 Sensitivity of GHG Impacts from Variations in Biofuel GHG Emission Coefficients: *Removing all Provisions* Scenario (All quantities are changes from baseline, million t CO$_2$-e, 2014-2021 annual average.)

	EISA	EPA w/ILUC	EPA w/o ILUC	High Biofuel Emissions	Low Biofuel Emissions
With RFS2	-5.4	-2.2	+3.5	-14.9	+6.7
No RFS2	-7.0	-5.6	+2.4	-14.1	+3.7

capture impacts within the domestic U.S. energy sector—between sectors (e.g., transportation, industrial, commercial, and residential) and between energy sources that could be caused by the effect of the biofuel provisions and the inter-acting RFS on relative fuel prices within the United States. Moreover, NEMS-NAS has a longer time horizon (out to 2035) than FAPRI-MU (one decade), which allows insight into the temporal dynamics induced by the provisions.

Figure 5-1 presents NEMS-NAS results on the effect of the biofuel provisions on the composition of biofuel production (use). The reference scenario is simply the baseline described above. "No Bio Subsidies" is equivalent to the "Remove All Provisions" scenario above.

The results outlined in Figure 5-1 can be summarized as follows:

- The RFS continues to motivate biofuel production, even in the absence of the biofuel credits, though at slightly lower levels when the provisions are removed.
- Corn ethanol production decreases when the subsidies are removed because ethanol's value is reduced.

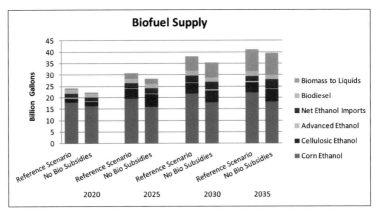

FIGURE 5-1 Effects of biofuel provisions removal on biofuel production levels.

- On the other hand, ethanol imports (primarily Brazilian sugarcane ethanol) increase when the import tariff is removed.
- Biodiesel decreases slightly, and the mix of cellulosic fuels produced shifts slightly as well with more cellulosic ethanol and lower biomass to liquids.
- NEMS-NAS does not separate out the effects of the provision on gasoline use in the United States or other countries.

Figure 5-2 reports the impacts of removing the biofuel tax provisions on total U.S. energy expenditures by sector. Higher transportation fuel prices mean higher household energy expenditures (maximum of 1.3 percent for transportation in 2027 or 0.3 percent overall).

Figure 5-3 shows the impact of removing the provisions on federal revenues, indicating:

- In the reference scenario, which projects the effects had the ethanol tax subsidies and import tariff stayed in force through 2035, the cost of the biofuel credits rises to roughly $18 billion per year while the ethanol import tariff brings in almost $0.8 to 1.1 billion per year.
- When the credits and tariff are removed, the gain to the Treasury is roughly $17 billion per year or $300 billion total from 2012 to 2035.

Figure 5-4 presents NEMS-NAS estimates of the effects of the biofuel provisions on total CO_2 emissions in the U.S. Key findings are:

- Removing the biofuel credits makes virtually no difference in CO_2 emissions, because the RFS mandate requires continued production of biofuels.

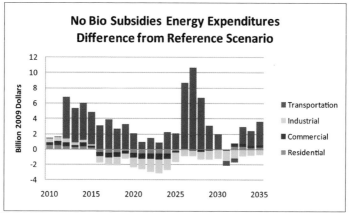

FIGURE 5-2 Effects of removing the biofuel provisions on total energy expenditures.

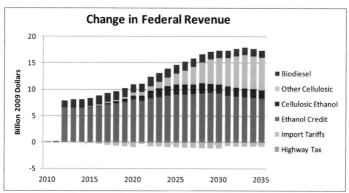

FIGURE 5-3 Effects of biofuel provisions removal on federal revenue.

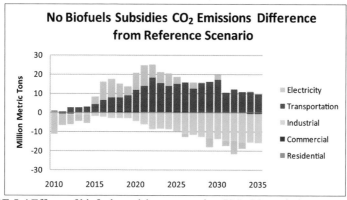

FIGURE 5-4 Effects of biofuel provisions removal on U.S. CO$_2$ emissions.

- Lower corn ethanol production and coal with biomass to liquids leads to lower CO_2 emissions in the industrial sector.
- Lower E85, lower biodiesel, and higher gasoline consumption leads to an increase in direct CO_2 emissions from the transportation sector.

One of the most important questions is to determine how the two models reviewed here compare in their projections. It should be emphasized that the modeling structures are highly differentiated (see Appendix A for a detailed review). Although the source of all emissions tracked by NEMS-NAS and FAPRI-MU differ in terms of sectors, types of gas, and countries covered, the essential finding is the same—the biofuels tax provisions have a very limited impact on GHG emissions. This result arises in part because the quantitative response on the fuel market is small and in part because the renewable fuel standards constrain the effects of tax changes on ethanol production.

SUMMARY

This chapter estimates the effects of biofuel tax provisions on global and domestic GHG emissions. While the biofuels provisions expired at the end of 2012, the committee included these in its list of provisions for several reasons: because ethanol credits have been widely used in the U.S. and abroad and raise important public policy questions, because the results illustrate the often-unintended impact of tax expenditures, and because the biofuels standards inter-act significantly with motor fuels taxes and the use of petroleum.

The effects of removing the Volumetric Ethanol Excise Tax Credit indi-vidually and all provisions together are tested using the FAPRI-MU model and the NEMS-NAS model. The findings from both models suggest the VEETC has very little effect on GHG emissions relative to the baseline. Removal of the VEETC in the FAPRI-MU simulation results in a reduction of emissions of 4.8 MMT CO_2-e per year during the first decade, at a value of 0.0007 tons CO_2-e per dollar of revenue.

Similarly, it is found that the remaining provisions have little effect on GHG emissions, with removal of all provisions resulting in a reduction of emis-sions of 5.4 MMT CO_2-e per year, at a value of 0.0004 tons CO_2-e per dollar of revenue (FAPRI-MU model).

The global market linkages affect the net impact of the provisions on global GHG emissions. Removal of the provisions resulting in a decrease in world ethanol use and an increase in world gasoline use.

It is also found that the RFS2 biofuels mandate—which requires a minimum quantity of biofuels to be mixed into motor fuel each year regardless of the subsi-dy—influences the impact of the biofuels provisions on global GHG emissions. Removing both the tax provisions and the RFS2 mandate results in a reduction of emissions of 7.0 MMT CO_2-e per year, at a value of 0.0007 tons CO_2-e per dollar of revenue. Therefore, as is intuitive, the mandates reduce (in absolute value) the size of the impact of the tax subsidy on GHGs. This effect—in which regulatory mandates limit the size of the impact of taxes and subsidies—is a common finding that runs through this report.

A key uncertainty in the estimates is the magnitude of indirect land-use change that is caused by the various biofuel feedstocks. However, the range of estimates of biofuels-induced ILUC from the literature suggests that increasing the use of certain forms of biofuel results in increased CO_2 emissions.

Taken together, the modeling results and existing literature suggest that, when the renewable fuel standards are in place, the biofuels provisions of the tax code have a small net effect on global GHG emissions. Although the effects are small, they are likely to increase GHG emissions slightly when key factors such as petroleum substitution and indirect land-use change are taken into account.

Chapter 6

Greenhouse Gas Emissions and Broad-Based Tax Expenditures

INTRODUCTION

Chapters 3 through 5 analyzed the effects targeted tax provisions have on greenhouse gases (GHGs). The current chapter examines broad-based tax expenditures. While the narrow provisions may cause significant impacts because they affect the energy sector, which has relatively high GHG emissions per unit of output, the broad-based provisions may be important because of their size and importance for the overall economy.

Broad-based provisions of the tax code are likely to have two kinds of impacts on GHG emissions. First, they may change the composition of national output from high (low) GHG-intensive sectors to low (high) GHG-intensive sectors. Second, they may increase or reduce the overall size of the economy and therefore change emissions simply because the economy is larger or smaller.

Although there are many large tax expenditures, here as in other areas the committee found it necessary to limit the scope of its inquiry. After examining the largest tax expenditures, the committee decided to examine three broad types of provisions: (1) tax incentives that affect investment generally, (2) tax incentives for owner-occupied housing, and (3) tax incentives for the provision of health insurance.

Among tax incentives for investment, the one that results in the largest revenue loss is the provision allowing firms to claim accelerated depreciation for investment in machinery and equipment under the Modified Accelerated Cost Recovery System (MACRS).[1] Tax incentives for housing include the mortgage

---

[1] Accelerated depreciation for machinery and equipment was the largest business tax expenditure in the lists compiled by the Office of Management and Budget (OMB) and Joint Committee on Taxation (JCT) in 2012. In 2013, however, JCT shows the 5-year revenue loss from the provision as negative. This reflects the fact that special temporary bonus depreciation in rules in effect in 2011 and 2012 expired, so that companies will be

interest deduction, the second largest income tax expenditure in the tax code, as well as the deduction for property taxes on owner-occupied homes and the partial exemption of capital gains on the sale of a principal residence. The single largest tax expenditure, according to both Joint Committee on Taxation (JCT) and Treasury Department estimates, is excluding employer-provided health insurance benefits from being taxed as income. Health insurance benefits are also excluded from wages used to determine payroll tax liability. We discuss each of these provisions in detail below.[2]

FINDINGS FROM PRIOR LITERATURE

More than the other provisions considered by the committee, the literature provided very little in the way of prior research on the impact of the tax code on GHG emissions. For example, a few studies consider the effects of the tax code's subsidies on housing, while other studies consider the impact of the capital gains exclusion on the rate of home sales. None of the studies the committee or consultants found consider the impacts of housing-related tax provisions on GHG emissions (Shan, 2011; Cunningham, 2008). Many studies estimate the effects of the subsidy provided by the mortgage interest deduction on federal receipts, the distribution of the tax burden, and housing prices, and its differential effects across states and among metropolitan areas, but none address its potential effects on GHG emissions. Important studies of these provisions include Poterba 1992, 2011; Toder, 2010; Gale, 2007; Glaeser, 2003; Sinai, 2001; Mann, 2000; Brinner, 1995; Capozza, 1996; Voith, 1999; and Bruce, 2001.

Research on the provisions for accelerated depreciation considers the impact on the relationship between capital and energy but makes no attempt to assess greenhouse gas impacts (Solow, 1987; Field, 1980; Metcalf, 2008). While there is a voluminous literature on the economic impacts of the tax expenditures for health care, the committee's review of the literature found no studies addressing energy or greenhouse gas impacts of the tax subsidies to health care.

---

receiving much smaller deductions over the next few years from investments they made in 2011 and 2012 than would normally be the case. The tax expenditure figures show cash flow revenue effects, not the incentive effects of provisions.

[2]Some economists believe that the ideal tax system is a consumption tax, which exempts the return to saving, instead of an income tax. Under a consumption tax, the mortgage interest deduction and the exclusion of employer-provided health insurance and health care would still be tax expenditures, but accelerated depreciation would not be. Furthermore, under a consumption tax, the normal rule would be to allow capital costs to be deducted in the year incurred, so that any depreciation system that spreads deductions over time would be viewed as a tax penalty. Such a system, however, would also have many other features that differ from the current law, including disallowing deductions for interest expenses. Given that the United States in fact has an income tax, the committee viewed the tax expenditure definitions used by the Treasury and the JCT as the appropriate ones to use for this study.

MODELING APPROACH

The committee undertook a literature review to determine whether the existing literature provides estimates of the impact of broad-based tax expenditures on GHG emissions. The answer was negative. While there is an extensive literature on the economic impacts of tax expenditures (see, for example, Poterba, 2011), no complete study has addressed the questions posed to the committee.

Given the lack of existing work, the committee decided here, as in the other areas, to undertake a study using existing models. After a review of possibilities, the committee concluded that the most promising model to simulate the effects of these broad-based tax policies on GHG emissions was the Intertemporal General Equilibrium Model (IGEM), maintained by Dale Jorgenson Associates (DJA).

IGEM is an integrated energy-economic model of the U.S. economy that includes the following features that are useful for this investigation: (1) a detailed representation of the U.S. tax structure, (2) a detailed breakdown by industry sector of output in the United States economy, including representation of industries involved in the production and consumption of energy, and (3) a disaggregated representation of the household sector based on data for individual households. Because of its structure, IGEM can provide many useful insights about the effects of these policies on GHG emissions. Nonetheless, IGEM, as discussed below, also has certain limitations for analyzing some of the broader tax policies discussed in this chapter.

A more complete description of the IGEM model and alternatives is contained in Chapter 2 and in Appendix A. One important feature leads to a major difference between IGEM and other models that have a detailed representation of the energy sector, such as the National Energy Modeling System (NEMS model). IGEM does not have an explicit representation of different types of capital in the energy sector, or of the details of energy supply. In conjunction with the putty-putty nature of capital (see further discussion later in this chapter and in Appendix A), this may lead to quite different responses in the energy sector to price changes. For example, a tax policy might lead to an increase in the demand for electricity. In the NEMS model, this might be met at the outset from existing capacity, up to a point. But beyond that point, the NEMS model would recognize the long lead times of nuclear power, the tightening environmental restrictions on coal, and the current advantageous position for these reasons of natural gas-fired generation. The IGEM assumptions, by contrast, would simply switch all generation to the cheapest power source almost immediately. Because fossil electrical generation has a much higher GHG intensity than other fuels and renewable sources, IGEM is likely to (and in some cases we review below, probably does) project some tax policies as having greater GHG impacts than other models would suggest.

The ground rules for the modeling of broad-based tax expenditures were similar to those described in Chapter 2. The model calculated the impact of removing tax provisions on the economy and on GHG emissions. Because IGEM is a general equilibrium model and explicitly represents federal government revenues and expenditures, it was necessary to specify the fiscal adjustments. The simulations were performed keeping the budget deficit unchanged on a year-by-year basis. This was done by returning any revenues using two alternative assumptions about how the revenue from eliminating tax expenditure provisions was recycled: first, as lump-sum payments to households, and second, as an equal proportional reduction in all marginal tax rates applied to individual and corporate income.

The background data on GHG intensities will be useful in understanding the current structure of the economy. Table 6-1 shows IGEM's estimated carbon dioxide (CO_2) emission intensities of different sectors for the year 2010; these coefficients are based on input-output use tables. We show the five most CO_2-intensive sectors and the five least CO_2-intensive sectors. Additionally, the sectors considered in this study are shown in bold. An important consideration in thinking about tax policies is how very intensive a few sectors are compared with the rest. For example, electric utilities are approximately 20 times more CO_2-intensive that the economy on average. This implies that small shifts in the composition of output can have large impacts on overall GHG intensity because of changes in the composition of output within the economy.

TABLE 6-1 Energy Intensities of Different Sectors

Intensities in CO_2-e per unit real output	
Five most intensive	Emissions intensities
Electric utilities (services)	5.740
Petroleum refining	4.412
Gas utilities (services)	2.746
Primary metals	2.293
Coal mining	1.887
Personal and business services	0.108
Five least intensive	
Apparel and other textile products	0.055
Finance, insurance and real estate	0.034
Non-electrical machinery	0.027
Communications	0.018
Electrical machinery	0.015
Consumption	0.168
Investment	0.030
Government	0.106
Exports	0.335
Total	**0.247**

Note: Emissions in millions of metric tonnes CO_2 equivalent ($MMTCO_2$-e). Domestic industry output and GDP quantity indices in $1996 billions. Source: IGEM model documentation. Source: IGEM model documentation.

Tables 6-2 and 6-3 present the major results of the IGEM analyses. Table 6-2 shows the basic results of the IGEM calculations. The striking result is that the major impact, if there is one, comes through the output effect. For all the broad-based tax expenditure scenarios, GHG intensities change by less than 0.4 percent from the base case when the tax expenditure provision is removed. But in all cases except for accelerated depreciation, removing the tax preference and substituting lower tax rates raises total emissions, because the lower tax rates on labor and capital income raise national output. In the case of removal of the accelerated depreciation preference, national output also increases, because the capital stock, although smaller, is deployed more efficiently and lower tax rates raise saving and labor supply. However, the decline in total emissions per unit of output from lower capital investment reduces GHG emissions by more than the higher output increases them.

In the following sections, we describe the results of the modeling in more detail along with an interpretation of the results.

INVESTMENT INCENTIVES

Investment incentives encourage capital formation and tilt the playing field towards business investment in tax-favored assets. Both the shift towards more capital-intensive production methods and the reallocation of output among industries can affect the level of GHG emissions, even for incentives that do not directly target energy production or conservation investments. This section discusses how accelerated depreciation for machinery and equipment—a long-standing preference in the federal income tax—may affect both the total level of GHG emissions and emissions per dollar of national output.[3]

---

[3]This report uses the term "national output" as an indicator of overall income and output. In principle, it would include improvements in the allocation of resources that are difficult to capture in the national economic accounts. The committee prefers the net national product (NNP) as its measure of national output and income to gross domestic product (GDP). NNP is a better measure of a nation's output than the standard measure. NNP differs from GDP in two respects. First, it subtracts depreciation of capital from the measures of investment and total national product. By so doing, it measures the net addition to the nation's capital stock (and therefore future living standards) instead of total spending on new investment (gross investment), including investment that simply replaces capital that has worn out. Second, it counts as investment amounts invested by U.S. residents overseas as well as abroad, whereas GDP measures only domestic investment. This second correction measures net additions to wealth owned by U.S. residents instead of additions to capital in the United States, some of which will provide future income to foreign residents. IGEM assumes that the U.S. current and financial account balances of the balance of payments are unaffected by policy changes, so for IGEM the only difference between changes in NNP and changes in GDP is changes in depreciation of capital. By using the NNP measure, we recognize that higher depreciation does not contribute to living standards.

TABLE 6-2 Effects of Different Revenue Recycling Options (% change from reference scenario, average over 2010-2025)

	EPHI: Tax on Services	Mortgage interest deduction	Property tax deduction	Capital gains on homes	Accel. Depreciation	Petroleum Tax
Tax rate cut case						
GDP	1.03	2.25	0.62	0.26	0.20	-0.16
Total GHG	0.65	2.35	0.68	0.22	-0.17	0.99
CO_2	0.65	2.28	0.66	0.21	-0.11	1.28
GHG intensity (per unit GDP)	-0.38	0.10	0.06	-0.04	-0.33	1.22
Lump sum rebate case						
GDP	0.45	-0.04	-0.09	0.21	-1.87	0.11
Total GHG	-0.74	0.23	0.02	0.17	-2.06	1.24
CO_2	-0.71	0.20	0.01	0.17	-1.98	1.53
GHG intensity (per unit GDP)	-0.29	0.28	0.11	-0.04	-0.17	1.20

Source: Dale Jorgenson Associates, "Effects of Provisions in the Internal Revenue Code on Greenhouse Gas Emissions," Report to the Board on Science, Technology and Economic Policy of the National Academies, June 6, 2012.

TABLE 6-3 Estimated Effects Relative to Base Case of Eliminating Accelerated Depreciation Preference on Key Economic Variables, 2010-2035

Economic Variable	With Lump-sum Rebate (%)	With Cut-in Marginal Tax Rates (%)
National Output (NNP)	-1.65	0.38
Capital Stock	-2.54	-0.50
Labor Input	-0.81	1.01
Coal Mining	-2.93	-0.42
Electric Utilities	-2.66	-0.64
Total GHG Emissions	-2.06	-0.17
CO_2 Emissions	-1.98	-0.11
GHG Emission Intensity	-0.34	-0.49

Source: Dale Jorgenson Associates, "Effects of Provisions in the Internal Revenue Code on Greenhouse Gas Emissions," Report to the Board on Science, Technology and Economic Policy of the National Academies, June 6, 2012.

Description of Provisions

The federal income tax includes a number of tax expenditures to encourage private investment. The largest of these is the provision to allow accelerated depreciation of machinery and equipment. Under current law, the costs of investment in machinery and equipment can be recovered using asset lives and depreciation methods specified under the Modified Accelerated Cost Recovery System, a system that allows companies to deduct the cost of assets at rates faster than the rates at which their values actually decline. In addition, in certain years, companies have been allowed to claim bonus depreciation allowances, which allow them to deduct a portion of the cost in the year the investment is made, with the remaining amount then recovered using MACRS. In 2011, for example, companies could recover 100 percent of the cost for machinery and equipment immediately (this being expensing), while in 2012 the bonus depreciation rate was 50 percent.

The Office of Management and Budget (OMB) estimated that accelerated depreciation of machinery and equipment will reduce federal revenues by $354.1 billion between fiscal years 2011 and 2015 and by $374.6 billion between fiscal years 2013 and 2017.[4] The Joint Committee on Taxation (JCT) estimated the cost at $109.0 billion between fiscal years 2011 and 2015. OMB describes its estimate of the tax expenditure as the difference between revenue under existing tax provisions and a hypothetical alternative that allows businesses to claim depreciation over time in accordance with the decline of the economic value of assets due to wear, tear, and obsolescence. JCT describes the tax expenditure as "depreciation of equipment in excess of the alternative depreciation system." Because the two agencies do not necessarily measure the tax expenditure against the same baseline law and may make different assumptions about levels and timing of investment in qualified equipment, they arrive at very different estimates of the cost of the tax preference.

Economic Effects

Accelerated depreciation reduces the cost of investing in machinery and equipment and thereby raises the level of investment in the short run and the capital intensity of the economy in the long run. Higher investment raises the growth of real output in the short run and raises output per capita and real wages in the long run by increasing capital per worker.

But the net economic benefit of accelerated depreciation depends on how it is financed and what other changes take place. Higher fiscal deficits could raise interest rates and crowd out investment, while paying for the tax benefit through raising other tax rates or cutting spending could have other potentially harmful effects, depending on how taxes are raised or what types of spending are reduced.

There are a number of channels through which accelerated depreciation could affect greenhouse gas emissions:

---

[4]The Office of Tax Analysis at the U.S. Treasury Department performs the estimates for OMB for inclusion in each year's federal budget documents.

- Positive effects on economic growth could raise GHG emissions, simply because a larger economy, with all other things unchanged, will require greater use of fossil fuels.
- A more capital-intensive economy could raise or lower GHG emissions per unit of output, depending on whether capital and energy are substitutes or complements in production.
- Shifts in output among industries (from those that use less to those that use more machinery and equipment) could increase or decrease fossil fuel use, thereby raising or lowering GHG emissions per unit of output.
- Accelerated replacement of old capital could reduce GHG emissions per unit of output if new capital is more energy efficient than the capital it replaces.

The combination of these effects means that eliminating the accelerated depreciation preference could either increase or decrease GHG emissions. Additionally, policy makers will also be interested in knowing the extent to which changes in GHG emissions reflect changes in the level of national output (all things the same, higher output will be associated with more GHG emissions); the extent to which they reflect changes in the composition of output; or the possibility that changes affect the emissions intensity of different sectors.

Assessment of Modeling Results

IGEM represents the accelerated provisions in current law as based on the MACRS provisions that are permanently in the tax code, assuming that the temporary bonus depreciation provisions will expire as scheduled. The effects of eliminating MACRS are estimated by assuming that it would be replaced by capital cost recovery rules consistent with economic depreciation.[5] This would raise the cost of capital in different industries by amounts that vary with the composition of assets (machinery and equipment, structures, and inventories) used in the industry and the degree of preference currently given to the equipment specific to the industry by MACRS. IGEM is well designed to capture these effects.

Some features of IGEM may overstate the effects of reducing the cost of capital on national output and growth. One important feature is the assumed flexibility of the capital stock (this is called the putty-putty assumption). The

---

[5]The report submitted by Dale Jorgenson Associates (2012) provides the formula they used for computing the cost of capital under economic depreciation, but does not specify how they derived the economic depreciation rates for the various industries. One source for calculating these parameters is the asset life categories used by the Bureau of Economic Analysis (BEA). The BEA asset lives can be combined with assumptions about the pattern of decline in asset values to compute estimates of the annual percentage depreciation rate.

structure of capital is assumed to respond immediately to changes in prices. This is particularly questionable for long-lived assets in the energy sector, such as nuclear power plants. A more realistic assumption, which is followed, for example, in the NEMS model, is that once capital is in place, its characteristics and particularly its input mix are fixed (this is called the putty-clay assumption). While the reality is somewhere in between these polar cases, it is clearly unrealistic to assume that a coal-fired plant can overnight turn into a wind plant if coal prices rise sharply or if the regulatory environment changes quickly.

IGEM assumes that agents (households and firms) have an infinite time horizon. This can be interpreted as assuming the current generation acts as if the well-being of their offspring is perfectly substitutable for their own well-being. Some other general equilibrium models, in contrast, are based on the life-cycle hypothesis that households allocate consumption over their lifetime to optimize their own economic well-being, given their lifetime resources (initial wealth and present value of lifetime earnings). The effect of the infinite horizon assumption is generally to increase the effect of investment tax preferences on current investment. Taken with other assumptions in IGEM, the result is that after-tax interest rates are unaffected by tax and other policy changes, so that increased demand for tax-favored investments will not drive up interest rates and crowd out other investment.

IGEM is a model with fixed current account and financial account flows in the balance of payments. This implies that none of the adjustments to saving and investment occur outside the United States. This will overestimate the impact of changes on domestic consumption and saving, which is assumed to respond completely to changes in investment demand. In a broader and more accurate model that allows for changes in international capital flows, changes in the demand for investment could also result in changes in international capital flows, as saving from abroad finances some of the new investment.

Table 6-3 shows the estimated effects of replacing the current accelerated depreciation preference (MACRS) with economic depreciation. In the case where the increased revenues are used to provide lump-sum rebates to households, national output would decline an average of 1.65 percent per year over the 26-year period 2010–2035 (Table 6-3). Because lump-sum rebates leave the tax structure unchanged, this simulation is the closest to isolating the effects of the tax preference alone. It indicates that accelerated depreciation provides a significant boost to national output, largely due to an increased capital stock. The capital stock will eventually be 2.5 percent smaller than in the base case scenario if the preference were removed. Labor input would also decline, as the reduced capital per worker lowers real wages (compared with the baseline) and therefore causes households to substitute leisure for labor (i.e., work less).

Total GHG emissions and CO_2 emissions both decline about 2 percent relative to the baseline, reflecting to a large degree reduced investment and output in the coal-mining and electric-utility sectors. The decline in emissions is larger than the decline in national output, so the emissions intensity of production falls by about 0.34 percent. Put differently, if accelerated depreciation provisions are removed, GDP declines 1.65 percent, emissions intensity falls by 0.34 percent, so emissions decline by 2.06 percent.

The second column of Table 6-3 shows the impact when revenues are raised through tax-rate changes. When the tax revenue from eliminating accelerated depreciation is used to lower all marginal tax rates on individuals and corporations, national output increases by about 0.4 percent, relative to the baseline projection. In other words, the impact on national output is close to neutral if the capital tax-expenditure provisions are replaced by tax-rate reductions. The capital stock still falls (by 0.50 percent), but the reduction in marginal tax rates induces a 1.01 percent increase in labor supply. The reduction in preferential treatment of those selected industries that benefit most from accelerated depreciation leads to a more efficient allocation of the capital stock.

The impact of removing accelerated depreciation on total emissions of GHGs is essentially zero in the case when revenues are recycled through lower tax rates. While national output is higher by about 0.38 percent, the emissions intensity of the economy declines by 0.49 percent. The net effect on GHG emissions of -0.17 percent is essentially zero and probably not within the resolution of the model.

Conclusions: Accelerated Depreciation

Accelerated depreciation is one of the largest tax expenditures in the federal income tax code (although, as indicated above, the cost of the preference is imprecisely estimated). According to the estimates prepared for the committee, its overall impact on national output is uncertain and depends upon the method by which the revenues are recycled. Eliminating accelerated depreciation would in both recycling cases reduce the capital stock. However, the effect on capital stock would be partially offset and labor supply and national output would increase if the additional revenue were used to finance cuts in individual and corporate marginal tax rates.

The impact of removing accelerated depreciation on overall GHG emissions is probably negative, but the amount depends upon the fate of the revenues. If the revenues are returned by lowering tax rates, then the overall impact on GHGs is essentially zero. In contrast, if they are returned through lump-sum rebates, then GHGs are probably lower because the lower emissions intensity is combined with lower economic growth, and overall emissions are calculated to fall by about 2 percent.

TAX INCENTIVES FOR OWNER-OCCUPIED HOUSING

Federal tax incentives for owner-occupied housing are usually justified as encouraging more people to own homes. Homeownership is often held to have positive economic spillovers, such as better property maintenance, and higher levels of community involvement and voting participation. But the incentives also encourage people to own larger houses, affect patterns of urban development, impact the rate of housing stock turnover, and reallocate capital from the business to the household sector. Each of these changes could affect greenhouse gas emissions, and possibly in opposing directions.

Description of Provisions

The federal income tax provides significant tax incentives for investments in owner-occupied housing. First, although investment income actually received is generally taxable under the U.S. federal income tax, implicit investment returns are generally not taxable. In the present case, the implicit income homeowners receive from housing services they provide to themselves is exempt from taxation. This implicit return is called "imputed rent on owner-occupied housing" and represents the net rental income that owners would receive, after deducting costs of operation, maintenance, depreciation, and interest, if they had to pay rent to themselves as tenants. JCT and the Treasury Department originally did not include the exemption of net imputed rent as a tax expenditure provision on the grounds that it would be impractical to require homeowners to place an implicit value on the rent they effectively pay to themselves and report that rent as income.[6] In recent years, Treasury, but not JCT, has listed imputed rent as a tax expenditure line item. Treasury estimates that the exclusion of net rental income will cost the government $337.4 billion in lost revenues between fiscal years 2013 and 2017. The committee chose not to include the tax preference to imputed rent in our simulations, primarily because none of the major tax-reform proposals include this on the list and it is in practice difficult to define and enforce.

Capital gains from the sale of owner-occupied homes also receive favorable tax treatment, compared with how the tax law generally treats capital gains. Homeowners may exclude from gross income up to $250,000 ($500,000 for a married couple filing a joint return) of the gain from the sale of a principal residence if the taxpayer has owned and used the property for at least 2 of the 5 years preceding the date of sale. OMB scores the cost of the capital gains exclusion as $171.1 billion during fiscal years 2013 and 2017 and $121.1 billion during 2011 and 2015. JCT estimates the cost at $123.2 billion during fiscal years 2011 and 2015.

In general, business firms and investors can deduct from their taxable income costs of investment, such as interest paid and state and local property taxes. But because owner-occupied homes do not generate taxable income (due to the exclusion of imputed rent), deductions for housing costs are regarded as tax

---

[6]Although not regarded as part of their income by most people, imputed rent is counted as income in the U.S. National Income and Product Accounts on the grounds that the measured capital return generated by the nation's housing stock should not change when people switch from being tenants to owners.

preferences.[7] OMB estimates that the deduction of mortgage interest on owner-occupied housing will cost $606.4 billion during fiscal years 2013 and 2017, making this deduction the second largest tax expenditure in the federal income tax. Between fiscal years 2011 and 2015, OMB scores the cost of this provision at $491.1 billion, while JCT scores its cost at $464.1 billion. OMB estimates that the deduction of state and local property taxes on owner-occupied homes will cost $140.6 billion between fiscal years 2013 and 2017 and $118.6 billion between 2011 and 2015. In comparison, JCT estimates the cost at $117.1 billion between 2011 and 2015.

Economic Effects

Most taxpayers receive little or no benefit from the tax preferences for owner-occupied housing other than the exclusion of imputed rent. Taxpayers who do not itemize deductions on their tax returns receive no benefit from the deductibility of mortgage interest or state and local property taxes.[8] Taxpayers with modest incomes who are in the 15 percent rate bracket receive a relatively small subsidy, compared with the benefit received by higher-income taxpayers in the 28 percent or 35 percent rate bracket.

JCT reports that the 18 percent of tax returns with incomes of $100,000 and over received 78 percent of the benefits from the mortgage interest deduction and 73 percent of the benefits from the real estate tax deduction. Among beneficiaries of the mortgage interest, 55 percent had incomes of $100,000 or more, while 52 percent of those claiming deduction for residential property taxes had incomes of $100,000 or more.

Because the mortgage interest and property tax deductions largely benefit those with high incomes who are likely to have very high home ownership rates without a tax preference and does little or nothing for lower income families, it probably does little to increase the rate of homeownership.[9] It does, however, substantially reduce the costs of housing capital. This could lead either to the construction of larger and more expensive homes, or to the bidding up of prices in areas, such as densely populated urban centers, where there is little space for housing expansion.

The JCT does not report the distributional effects of the tax preference for housing capital gains, but it is probably also highly concentrated at the top end of the income distribution, even though it is capped for taxpayers with very large

---

[7]Interest incurred to finance the acquisition of other non-income-producing household assets, such as cars, TV sets, and furniture, is not deductible.

[8]JCT reports that, in 2010, only 29 percent of taxpayers claimed itemized deductions. But 72 percent of returns with incomes between $100,000 and $200,000 and 94 percent of returns with incomes of $200,000 and over were itemizers.

[9]See, for example, William G. Gale, Jonathan Gruber, and Seth Stephens-Davidowitz, "Encouraging Homeownership Through the Tax Code," *Tax Notes*, June 16, 2007.

housing gains. Taxpayers in the 15 percent bracket or below—about 75 percent of all returns—currently face a zero tax rate on capital gains, so the additional exclusion for housing gains does not benefit them. Instead, it benefits higher-bracket taxpayers who would otherwise face rates ranging from 15 to 23.8 percent on the first $250,000 ($500,000 if married) of gains from the sale of their house.

The overall consequence is that all the housing incentives probably have little effect on the decision whether to rent or to own a home, but they very likely do affect the allocation of capital between housing and other assets. By so doing, they probably reduce the productivity of the capital stock and lower national output relative to a more neutral tax system that treats different capital assets more similarly.

Effects on Greenhouse Gas Emissions

The mortgage interest deduction encourages a shift towards additional investment in owner-occupied housing at the expense of business-sector investment. This reallocation of investment reduces the overall efficiency of the capital stock, thereby lowering national output and reducing GHG emissions associated with higher production.

Beyond this, for any given level of national output, the reallocation of capital from business to housing capital has several effects on GHG emissions, including:

- The shift from other assets to housing could raise or lower GHG emissions if other industries generate less or more emissions per unit of output than housing. In particular, to the extent a shift from business assets is associated with a reduction in production of fossil fuels and in the generation of electric power, it could lower GHG emissions per unit of national output.
- The increase in the housing stock could raise GHG emissions per unit of national output if larger houses are associated with higher consumption of fossil fuels than the output that is replaced.
- The tax incentives for housing could also raise GHG emissions per unit of national output, if they contribute to reduced density of residences, raising commuting times and distances and leading to increased automobile use.

Thus, while the tax incentives for housing do not affect fossil fuel consumption and GHG emissions directly, they could have substantial indirect effects through changes in the composition of output between industries and changes in the composition and geographic dispersion of the housing stock.

Assessment of Modeling Results

IGEM is able to capture the effects on national output from provisions that favor residential capital over business-sector capital. It also captures the effects on national output and its composition from offsetting cuts in marginal tax rates that may accompany any reduction in housing tax preference as well as the effects of shifts in the composition of output that result from differences in energy intensity among industries (including the household sector).

Unfortunately, IGEM omits two potentially important effects of tax incentives for housing on GHG emissions: (1) the effects of changes in housing capital and spending on household energy use, particularly on the energy used in heating and cooling; and (2) the effects of changes in housing capital and spending on transportation that results from changing the pattern of dwelling density and location. Because of these omissions, the committee believes that the IGEM results do not capture completely the impact of the housing provisions on GHG emissions.

IGEM does not capture the effects of changes in the size and composition of housing units because IGEM represents housing as capital that uses energy in construction (including replacement investment for the capital that depreciates) and maintenance, but not in its operation. This means that the portion of annual output that is used to maintain and improve the stock of housing capital generates GHG emissions, but the portion of annual output that represents the service flow from the existing asset does not. To capture the full effects of increases in the housing stock on energy use, a model would have to represent the flow of annual consumption from the housing stock as a service that is produced with inputs from other sectors (electric power for air conditioning and lighting as well as oil, gas, or electricity for space heating). That is, a model would need to capture the complementarity between housing space and fuel consumption in the form of electricity use (for cooling and other uses) and direct fuel consumption (for heating), instead of simply representing housing space as a final consumption good that substitutes for other goods that use electric power generation and fossil fuels as inputs in production.

IGEM also does not capture any effects of the tax incentives on the spatial allocation of the housing stock. Therefore, it cannot capture any possible linkages between changes in the size composition of housing and changes in commuting patterns, the transportation of goods, and the like.

A final concern is that IGEM has low resolution of the housing sector. Housing is included in a broad sector, "Finance, insurance and real estate." Moreover, there is no distinction between owner-occupied and rental housing. The low resolution makes interpretation of the results particularly difficult.

Modeling Results

Table 6-4 shows the estimated impacts of changes in housing tax preferences according to IGEM. The simulation finds that eliminating the mortgage interest and property tax deductions would have almost no effect on national output if the tax revenue is recycled through lump-sum transfers to households, for example, through tax rebates or credits. The efficiency loss from capital misallocation totally offsets any output increase from lower taxes on capital. National output would increase slightly in spite of declines in both labor input and investment.

By contrast, using the revenues from eliminating the deductions to lower marginal tax rates on individual and corporate income would lead to substantial increases in the capital stock, labor input, and national output. The efficiency gain from an improved allocation of the capital stock causes national output to increase by about twice the increase in productive inputs.

The simulations also show that eliminating the tax subsidy on mortgage interest would raise GHG emissions per unit of output. For both recycling cases, removing the subsidy increases the GHG intensity of national output. This reflects compositional changes in the economy, as elimination of the subsidy increases the output of heavy energy-using sectors (coal mining and electric utilities) at the expense of sectors not using energy (services from housing capital). Overall GHG emissions are estimated to increase in both recycling cases, primarily because of the increase in national output.

TABLE 6-4 Estimated Effects of Eliminating the Home Mortgage Interest Deduction Relative to Reference Scenario, 2010-2035

Economic Variable	With Lump-sum Rebate (%)	With Cut-in Marginal Tax Rates (%)
National Output (NNP)	+0.17	+2.42
Capital Stock	-1.12	+1.13
Labor Input	-0.77	+1.22
Coal Mining	-0.11	+2.70
Electric Utilities	-0.55	+2.81
Total GHG Emissions	+0.23	+2.35
CO_2 Emissions	+0.20	+2.28
GHG Emission Intensity	+0.28	+0.10

Source: Dale Jorgenson Associates, "Effects of Provisions in the Internal Revenue Code on Greenhouse Gas Emissions," Report to the Board on Science, Technology and Economic Policy of the National Academies, June 6, 2012.

Overall, it appears that because of omission of major complements to housing, the modeling results do not fully capture the impact of changing housing subsidies on GHG emissions.

Conclusions: Housing Subsidies

According to IGEM estimates, eliminating the tax subsidies for owner-occupied housing and using the revenue to lower marginal tax rates would improve the efficiency of allocation of the capital stock and increase national output. According to the modeling results, the impact on overall GHG emissions would be determined primarily by the overall economic reaction: If the provision increases GDP, then GHG emissions would change at about the same rate.

However, the simulation does not accurately capture the full effects of the housing subsidy on emissions intensity. The simulation fails to capture the effects of housing size on household consumption of electricity, natural gas, and home heating oil and the potential effects of changing residential patterns on automobile use and gasoline consumption. Moreover, the resolution of the housing industry is low, with no distinction between rental and owner-occupied dwellings. On the whole, the committee finds that the simulation results do not present a complete picture. The committee therefore concludes that the existing literature, as well as the results commissioned for the present study, is inconclusive regarding the impacts of eliminating housing tax incentives on GHG emissions. Understanding the full impacts remains an important topic for future research.

TAX SUBSIDIES FOR HEALTH INSURANCE AND HEALTH CARE

As is the case for other broad-based provisions, tax subsidies for health care do not directly affect energy use. But removing them could lower or raise GHG emissions, depending upon the impact on national output and on whether the composition of the economy shifts to more or less GHG-intensive activities.

Description of Provisions

The largest single tax expenditure in the Internal Revenue Code is the exclusion of employer-provided health insurance (EPHI) benefits from taxable income of employees. OMB estimates that exclusion of EPHI benefits will reduce income tax revenue by $1,012 billion between fiscal years 2013 and 2017 and by $904.6 billion between fiscal years 2011 and 2015. JCT estimates the exclusion will reduce income tax receipts by $725 billion during fiscal years 2011 and 2015.[10]

The EPHI exclusion is treated as a tax-expenditure provision because a comprehensive definition of income holds that all compensation, whether in the form of cash wages or fringe benefits, is counted as taxable income. Generally, fringe benefits (such as the annual rental value of housing or automobiles supplied by employers) are taxable to employees, but there are a number of statutory exemptions. The most important of these exemptions is for health insurance and health benefits.

There are other tax preferences for health that IGEM could not simulate. These include the deduction of health insurance premiums for the self-employed, the itemized deduction for medical expenses in excess of 7.5 percent of adjusted gross income (10 percent under the alternative minimum tax, and, beginning in 2013 under the regular income tax as a provision under the Affordable Care Act), the deductibility of contributions to Medical Savings Accounts, and the exemption of investment income accrued within those accounts.

Economic Effects

The exemption of EPHI premiums provides an incentive for more employers to provide health insurance coverage to their employees, and for those employers who do provide coverage to provide more generous plans, with lower co-payments and deductibles and more types of health services covered. More generous insurance coverage, in turn, encourages employees to spend more on health care services. The exclusion may also reduce the *marginal* effective tax rate (METR) on earnings.[11]

Including previously exempt fringe benefits in taxable income, without any compensating reduction in other taxes, raises the average effective tax rate on earnings. If the base broadening is accompanied, however, by a revenue-

---

[10]EPHI benefits are also excluded from the wage base for payroll taxes used to fund Social Security retirement and disability benefits and Medicare Hospital Insurance benefits. OMB estimates that the exclusion will reduce payroll tax receipts by $619.2 billion during fiscal years 2013 and 2017, over 60 percent of the loss in income tax receipts. Some of the increase in the federal deficit is offset, however, by reduced Social Security retirement and disability benefits accrued by workers who will be credited with lower earnings used for benefit determination. These implications of EPHI benefits in the social insurance programs are excluded from the calculations for this report.

[11]The effect here is complicated and can be seen as follows. The average tax-effective tax rate on earnings can be expressed by the equation, $t_e = (t_s W + t_s aF)/(W+F)$, where t_e = the effective tax rate on earnings, t_s = the average statutory tax rate, W = cash wages, F = the cash value of fringe benefits and a = the share of fringe benefits included in the tax base. If fringe benefits are excluded from the tax base (a = 0), the effective tax rate is simply equal to $t_e = t_s W/(W+F)$. If fringe benefits are included in taxable compensation (a = 1), then the effective tax rate is equal to the statutory rate t_s.

neutral reduction in the statutory tax rate on earnings, then the effective tax rate on earnings will be unchanged. If—as with the present analysis—broadening the tax base is accompanied instead by a reduction in statutory marginal tax rates that is applied to both labor and capital income, the effective tax rate on earnings will rise and the effective tax rate on capital income will decline.[12]

The operation of the EPHI is particularly complicated, because the effect on tax rates differs for different workers. The average tax rate on earnings can affect the decision whether or not to work for pay (the extensive margin). But for those who are working, the decision to work additional hours (the intensive margin) depends on the *marginal* effective tax rate, that is, the additional tax paid per dollar of additional compensation.

Calculating how taxing EPHI benefits affects the METR on labor compensation requires an assumption about how EPHI benefits vary with additional hours of work. For many workers, the EPHI benefits are fixed or lump-sum – that is, their value is independent of wages and salaries. If EPHI benefits in any job are fixed as hours of work change, then taxing EPHI benefits raises the average effective tax rate on compensation, but leaves the METR unaffected. If in contrast, the value of EPHI benefits rises in proportion to the increase in wages, then eliminating the tax provisions that exclude EPHI from income tax increases the METR by the same amount as an equal revenue increase in the marginal tax rate on wages.

Generally, therefore, there is no simple way to characterize how taxing EPHI benefits changes the METR on labor compensation and the consequent incentive to work. At the extensive margin (deciding whether to work), it would reduce the incentive to work in the marketplace instead of working at home. Similarly, it may affect the decision to work full-time instead of part-time if, as is often the case, the availability of health benefits is restricted to employees who exceed some threshold level of hours worked. For workers considering small changes in hours worked (such as manufacturing workers taking on overtime), taxing EPHI benefits may have no effect on the incentive to work. These considerations imply that the taxation of EPHI benefits would potentially be important primarily for people who are on the work–no-work margin, and less important for prime-age workers with strong labor-force attachments. This suggests that the work incentives of EPHI provisions will be particularly important for secondary earners, for people near retirement age, and for workers who are potentially eligible for disability coverage.

---

[12]Strictly speaking, the effective tax rate on earnings may also be affected by the tax rate on capital income, which affects how much future consumption a worker can purchase with her current earnings. The assumption in this discussion is that IGEM does not incorporate this direct effect of capital income taxation on the incentive to work, but instead looks only at the effective tax rate on wages.

Assessment of Modeling Results

The committee asked for two separate pairs of simulations of the health care exclusion. For the first pair of simulations, DJA modeled the exclusion as a partial exemption from taxation of labor compensation (we call this "EPHI exclusion"). This set of simulations did not capture the effect of the exemption on the price of health care services and therefore did not capture any of the effects of shifting the composition of production between health care and other sectors.[13]

Consequently, in order to include sectoral shifts, the committee requested a second pair of simulations in which the elimination of the health care exclusion was modeled as if it were equivalent to imposing a new excise tax on the consumption of health services (we call this "EPHI services"). For each pair of simulations, as with the simulations of other incentives, the additional revenue from removal of the tax benefit was recycled in two different ways—as a lump-sum transfer to all households and as reductions in marginal tax rates proportionately on individual and corporate income.

We requested that IGEM simulate the elimination of the exclusion of EPHI from income tax as if it were a new excise tax because the reference scenario contains no tax preference for EPHI services. Imposing an excise tax is equivalent to removing a tax-expenditure subsidy from the sector. We note that the health sector, like the housing sector, is poorly resolved in IGEM. IGEM does not include health services as a separate sector but as part of a larger sector called "personal and business services." Consequently, the excise tax was imposed on the personal- and business-services sector in which health is contained. Health is a substantial part of this sector, and its GHG intensity is similar, so we might expect that this treatment is a reasonable approximation to a more detailed treatment.

Table 6-5 shows the results of the calculation. The change in tax treatment reduces national output (relative to the reference scenario) by 0.5 percent over the 26-year period (2010-2035) when the revenues are used to provide a lump-sum subsidy to households. In this case there are no efficiency gains to offset the efficiency costs of the new tax on EPHI services.[14]

---

[13]The EHPI exclusion has no effect on the marginal effective tax rate on labor compensation in IGEM and therefore no effect on the incentive to work. As a result, when EPHI benefits are made taxable and the revenue is used to provide lump-sum transfers to households, there is virtually no effect on the key variables that characterize the economy.

But when the revenue from taxing EPHI benefits is used to lower marginal tax rates on individual and corporate income, output, capital stock, and labor supply all increase by more than 2 percent relative to baseline projections.

[14]This is likely to be an anomalous result because IGEM does not start from the point where the service sector is subsidized (due to health care exclusion) so that the new ex-

By contrast, when the revenues from the excise tax are used to reduce marginal tax rates on individuals and corporations, the capital stock rises by slightly over 1 percent, national output rises by about 1 percent, and labor supply rises by slightly under 1 percent.

One of the surprising results of the simulations is that they show that GHG intensity (GHG emissions per unit of national output) declines in both simulations. With the lump-sum substitution, GHG emissions decline by more than the drop in national output, while with the offsetting cut in marginal tax rates, GHG emissions increase by less than the increase in output.

This result on GHG intensity appears anomalous because the business- and personal-services sector is less GHG intensive than the economy as a whole. As shown in Table 6-1, the GHG intensity of the sector is about 40 percent of the economy-wide GHG intensity.[15] We would therefore assume that reducing the size of the health sector would increase the size of the rest of the economy and thereby raise overall GHG intensity. This is not the case with the IGEM simulations.

TABLE 6-5 Estimated Effects Relative to Reference Scenario of Eliminating the Exclusion for Employer-supplied Health Insurance, 2010-2035

Economic Variable	With Lump-sum Rebate (%)	With Cut-in Marginal Tax Rates (%)
National Output (NNP)	-0.49	+0.98
Capital Stock	-0.26	+1.14
Labor Input	-0.51	+0.84
Coal Mining	-0.78	+1.03
Electric Utilities	-0.84	+0.61
Total GHG Emissions	-0.74	+0.65
CO_2 Emissions	-0.71	+0.65
GHG Emission Intensity	-0.36	-0.29

Source: Dale Jorgenson Associates, "Effects of Provisions in the Internal Revenue Code on Greenhouse Gas Emissions," Report to the Board on Science, Technology and Economic Policy of the National Academies, June 6, 2012.

---

cise tax may reduce a distortion instead of producing a new one. But it is perfectly consistent with the effects of imposing a new excise tax on an unsubsidized sector.

[15]In addition to the data shown in Table 6-1, the same results are found in the Department of Commerce's estimates of CO_2 intensity looking at the direct and indirect impacts using input-output techniques. See Department of Commerce (2009).

The committee discussed the result with the IGEM modelers and examined the underlying results for individual industries in IGEM to determine the source of the decline in GHG intensity. It appears that the decline in the output of services is accompanied by a decline in the production of the petroleum and utilities sectors. In other words, there is a decline in the share of the most energy intensive sectors with the tax on services. Moreover, because the GHG intensity of energy-intensive sectors is so high, even a small decrease in their share can produce a decrease in overall GHG intensity. The committee was unable to determine what economic mechanism was behind this shift, however, and the shift is not completely persuasive from an economic point of view. The committee concluded that this is likely to be a statistical anomaly and one that is not within the ability of IGEM to resolve accurately.

Conclusions: Health Subsidies

Simulations for the committee found that eliminating the EPHI exemption would raise national output if offset by cuts in marginal tax rates. The estimates probably underestimate the distortionary effect of the health subsidies, so the impact on national output would probably be larger if the model could provide a more detailed representation.

The impact in IGEM of removing the tax preferences on health on GHG intensity is anomalous. Economic intuition would suggest that eliminating health care subsidies would raise GHG emissions per unit of output because the health care sector is less energy-intensive than the rest of the economy (see Table 6-1). The modeling results from IGEM show the opposite effect, however, with a small decrease in GHG intensity. An examination of the simulations indicates that the results are driven by sectoral shifts that are not entirely plausible. The committee concludes that the model cannot resolve the impact of the health subsidies with sufficient precision are the impact on GHG intensity of removing the tax benefit for employer-sponsored health insurance is still an open question.

OVERALL CONCLUSIONS

This chapter examines the impact of broad-based tax expenditures on GHG emissions. The provisions examined were tax incentives that affect investment, tax incentives for owner-occupied housing, and tax incentives for provision of health insurance. The committee investigated the impacts of removing these provisions through runs commissions by a single model, IGEM.

The major finding for all three sets of provisions is that the impact on GHG emissions is primarily driven by the impact on the growth of national output. In most cases, the change in GHG emissions is close to the change in national output growth induced by the change in the tax provision, and there is little change in the emissions intensity (the emissions-output ratio). The primary result is therefore intuitively sensible. If the emissions-output ratio does not change markedly, then the change in emissions will be determined by the impact on national output.

A second finding is that the effects on national output and GHG emissions can be significantly influenced by the way the revenues are recycled. The committee examined two revenue recycling mechanisms, lump-sum rebates and tax rate reductions. The simulations find that the impact of each of the broad-based tax changes on output and GHG emissions could be up to 2 percent if the revenues are recycled through tax rate cuts that increase efficiency. They would be small (i.e., close to zero) if the revenues are recycled in a lump-sum fashion.

The third and parallel finding is that the impact of the broad-based provisions on emissions intensities is generally small. The impacts range from a -0.3 percent to a +0.2 percent change in emissions intensities over the period 2010-2035. The committee could not undertake a model comparison to test these findings, and no statistical tests of significance for the results were made with IGEM. Because of these limitations, the committee finds that no reliable estimates of the impacts of the provisions on emissions intensities can be determined on the basis of existing evidence, but the effect on emissions intensities is likely to be small. The summary result, therefore, is that changes in broad-based tax provisions are likely to have a small impact on overall GHG emissions outside of their impact on overall economic growth.

Finally, the impact of the broad-based provisions is extremely small relative to emissions growth over the period of investigation. IGEM projects a 43 percent change in emissions over the period in the reference scenario simulation. Even the tax preference with the largest impact would change the growth of emissions over the analysis period by at most a small fraction of the total emissions growth.

Chapter 7

Summary of Findings and Recommendations and Use of Tax Policy to Address Climate Change Policy

THE COMMITTEE'S CHARGE AND APPROACH TO THE STUDY

This report is the response of a committee appointed by the National Academies to a charge by the U.S. Congress to conduct "a comprehensive review of the Internal Revenue Code to identify the types of and specific tax provisions that have the largest effects on carbon and other greenhouse gas emissions and to estimate the magnitude of those effects." The committee is composed of experts in tax policy, energy and environmental modeling, economics, environmental law, and related areas.

The charge is extremely broad, encompassing a detailed federal tax code, a vast energy system, an evolving regulatory environment, and a complex network of interactions among different parts of the economy. In focusing its work, the committee decided to concentrate its efforts on four groups of tax code provisions and closely related spending and regulatory policies that have significant effects on the emissions of CO_2 and other greenhouse gases (GHGs).

1. *Energy-related tax expenditures.* Because the country's energy sector, dependent as it is on fossil fuels, is the largest source of greenhouse gas emissions, tax provisions specifically targeted toward energy are a logical place to start the assessment. Given time and budget constraints, the committee was unable to consider all of the energy-related provisions, but it did examine major ones such as the renewable electricity production tax credit and allowance for percentage depletion of petroleum and natural gas. These provisions are discussed in Chapter 3.

2. *Energy-related excise taxes.* There are a handful of excise taxes that affect GHG emissions, primarily taxes on gasoline and other motor fuels and taxes on air travel. These are discussed in Chapter 4.

3. *Biofuels provisions*. Tax subsidies for biofuels, particularly ethanol, interact with a complex set of mandates on renewable fuels, and the combined effects of these on energy prices and consumption depend on market interactions that include spillovers to global agriculture markets. These provisions are discussed in Chapter 5.

4. *Broad-based tax expenditures*. By far the largest tax expenditures are broad-based tax expenditures, including the exclusion of employer-provided health insurance benefits from taxable income, subsidies to owner-occupied housing, and accelerated depreciation of plant and equipment purchased by businesses. Although not targeted to the energy sector, they can affect both overall economic growth and the composition of output and thereby could potentially have a significant impact on U.S. greenhouse gas emissions. These are considered in Chapter 6.

The committee carried out its work in three stages. We first decided which provisions of the tax code to examine. Next, we reviewed existing research on the selected provisions to gain insights on the channels through which they could affect climate change and determine if prior estimates on impacts exist. Finally, we commissioned new analyses using several existing energy-economic models to estimate the impact of major provisions.

This chapter summarizes our principal findings with regard to the greenhouse gas impacts of the four sets of tax provisions. It then sets out the committee's recommendations, including both recommendations for improvements in the methods used in our analysis and our guidance for thinking about environmental tax policy.

CHOICE OF TAX PROVISIONS

As described in Chapter 1, we chose for close examination provisions of the tax code that were closely related to energy-intensive activities because the energy sector is the largest source of domestic GHG emissions.

The tax code provisions most clearly relevant to the committee's charge are excise taxes on energy consumption and energy-intensive activities, including taxes on motor fuels and taxes on air travel. A much larger set of provisions consists of those known as tax expenditures. Tax expenditures are defined by Congress as revenue losses attributable to provisions of the federal tax laws that allow a special exclusion, exemption, or deduction from gross income or that provide a special credit, a preferential rate of tax, or a deferral of tax liability. The U.S. Department of the Treasury (Treasury) and the congressional Joint Committee on Taxation (JCT) each compiles a list of provisions each year. The Treasury list for fiscal year 2014 provides estimates of the revenue impacts for 173 provisions, and the JCT also provides an independent list and set of revenue estimates.

Since it was not possible to analyze each of the provisions, the committee's first task was to select the most important ones in the context of the present study. For this purpose, we selected the provisions with the largest dollar benefits to the energy-producing sector. Additionally, we selected some of the large broad-based tax expenditures not targeted at the energy sector that might have substantial impacts on GHG emissions.

In the end, the committee analyzed excise provisions that account for 46 percent of all energy-related excise tax revenues as well as those that account for 71 percent of the calculated revenue loss from the 10 largest energy-related tax expenditures in 2011. The broad-based tax expenditures selected by the committee for analysis account for about one-third of the cost of all tax expenditures that year.

REVIEW OF EXISTING RESEARCH

The next step was to review existing research on the impact of the tax code on greenhouse gas emissions. This was undertaken by the committee, the staff, and a team of consultants hired specifically for this study.

One area with a substantial body of research pertains to the impact of gasoline taxes on fuel consumption and CO_2 emissions. Most of these studies are incomplete, however, because they do not incorporate important features of the U.S. tax and regulatory systems, such as biofuels taxes and subsidies, CAFE (Corporate Average Fuel Economy) standards, and regulatory mandates for ethanol and other biofuels. Moreover, few studies consider the impact of changing gasoline taxes on sectors outside the energy sector (the general-equilibrium effects) or the impacts of the changing revenue streams.

Although there is a large literature on tax expenditures, there is virtually no empirical research on the impacts of these provisions on GHG emissions. An exception is a recent study by the Center for Business and Economic Research (CBER) at the University of Nevada, Las Vegas, reviewed in Chapter 2. While the CBER study uses a simplified economic model, its results are useful as a benchmark to compare with the results from more detailed modeling studies.

Finally, we observe that there is a substantial recent literature that investigates approaches to reducing GHG emissions and achieving national and global climate-change objectives (such as limiting GHG concentrations or temperature increases) in the most efficient way. Both analytical and empirical studies of this type have concluded that the most efficient approach to emissions reductions is through uniform economy-wide taxes or regulations directly targeted on GHG emissions. A uniform carbon price creates incentives for consumers, producers, and innovators to adjust their activities so as to reduce emissions and encourage development of low-emissions technologies. A short discussion of this literature was provided in Chapter 2.

COMMISSIONED MODELING STUDIES

Committees of the National Research Council are generally asked to review and synthesize the existing scientific literature in a specified area. However, because of the limited existing research in the area of the committee's focus and the need to use a unified set of baseline assumptions to compare the effects of different tax provisions, the committee concluded that it would be necessary to commission new studies.

The committee determined that the most useful approach to analyzing tax provisions is the use of computable energy-economy models. It is virtually impossible to calculate the impacts of existing tax provisions on GHG emissions without computational simulations. Examining the effects of a particular tax provision over time entails projecting its impact on investment and consumption decisions, the impact of those decisions on energy production and consumption, and the resulting effects on GHG emissions. Analysis of each of these components requires using a formal model to simulate economic, revenue, and GHG outcomes under different policies.

Limited time and resources as well as best practice pointed to the use of existing integrated economic models. Moreover, models developed over many years represent best judgments of modeling teams about the historical paths, current state, and likely future paths of important variables as well as their response to external shocks or changes such as tax policies. There is simply no substitute for careful modeling for analyzing the problems the committee was asked to address.

The committee applied the following criteria in choosing among alternative models to be used for this report. First, the structure of the models needed to be sufficiently flexible so that the pertinent tax provisions could be introduced. Second, the models chosen should be widely used in other studies and subjected to peer review, and it will be helpful if they are familiar to researchers and policy analysts. Third, the analysis should be reproducible and the data on which it relies should be available to other investigators.

Fourth, the committee attempted when possible to obtain results from multiple models to provide a robustness check on the results. Fifth, none of the models should be under the primary operation or control of members of the committee. Sixth, the modelers were commissioned to perform the analyses only if they could run the models and deliver results to meet time and budget constraints.

As a final precaution, in advance of making firm contractual commitments with the modelers, we submitted the proposed protocol for the analysis along with the specific models and modeling assumptions to outside experts not affiliated with the committee. The experts were asked to analyze the protocol. They provided written comments, and the committee had a meeting to discuss their comments with the experts, and then responded to their critiques in writing.

Using these criteria, the committee selected four different modeling teams to undertake its analyses. Chapter 2 and Appendix A describe the models and the rationale for their selection. Like all economic models, the selected models have strengths and limitations. Two of the models employed (National Energy Modeling System for the National Academy of Sciences [NEMS-NAS] and Food and Agricultural Policy Research Institute at the University of Missouri [FAPRI-MU]) were detailed analyses of the energy and related sectors using a partial equilibrium (PE) framework. PE models describe the focal sectors in considerable detail, but do not fully capture its interactions with all other parts of the economy, a special concern because the energy sector is so large and plays such a key role in so much economic activity.

A third model (the Intertemporal General Equilibrium Model [IGEM]) encompassed the entire U.S. economy using a general equilibrium (GE) framework. GE modeling in principle solves some of the limitations of PE models, but it operates at such a high level of aggregation that it cannot capture many of the important features of the energy sector. IGEM in particular does not have highly resolved detail on choices among methods of electrical generation, nor does it capture the intricate regulations of the biofuels sector. The fourth model we employed (CBER) was a simplified PE energy supply and demand analysis that was specifically designed to estimate the GHG impacts of tax expenditures, but did so with a highly stylized modeling structure.

An important reservation about most analysis of the impact of tax provisions is that it pertains only to U.S. GHG emissions. Of the models used for this report, only the FAPRI-MU model estimated the global impacts of U.S. tax policy as they related to international agriculture, crop, and biofuel markets. Climate change is a global phenomenon that depends upon global emissions. While this shortcoming is primarily important for tradable goods like grains and oil, the committee notes this shortcoming with respect to the current modeling structures.

A second important concern is that current empirical models cannot reliably capture the impact of the tax code on innovation and technological change, or what is sometimes called induced innovation. A change in prices and output of a specific technology, such as solar photovoltaic or offshore wind, will generally lead to increased research and development (R&D), commercialization, and experience or "learning by doing." These will in turn lower the costs of production, the prices of products that use that technology, and spur further production. These considerations suggest that a subsidy may lead to increased R&D and learning by doing, and thereby promote specific technologies. Two of the models employed (IGEM and NEMS-NAS) included induced innovation in their structure. There is little consensus, however, on the mechanisms behind induced innovation or the magnitude of the response in the energy sector. For this reason, the committee did not attempt a separate analysis of this issue.

For each of the models, the committee specified a set of baseline assumptions on gross domestic product (GDP) growth, oil prices, the regulatory environment, as well as the tax system. The assumptions about U.S. GDP growth and path of world oil prices are derived from the Energy Information Administration's 2011 Energy Outlook (AEO11), which is widely used in the modeling community. The tax code and regulatory environment of 2011 chosen by the committee as the basis of its analysis were also included in the Energy Information Administration's 2011 baseline. In addition to the baseline, some of the provisions were analyzed using alternative assumptions for GDP growth and oil prices to provide sensitivity analyses. These alternatives were higher macroeconomic growth, low natural gas prices, and high petroleum prices.

In analyzing the tax provisions, the committee established a common set of tax and regulatory benchmarks for the analysis. The time frame, set to allow for the effect of investment decisions on GHG emissions, covered in most instances 2010-2035. We assumed a stable tax code in which the 2011 tax code provisions remain in place indefinitely. Note, therefore, that tax provisions that expired at the end of 2011 or 2012 are assumed to be extended indefinitely in our base case. This implies that the fiscal changes included in the American Taxpayer Relief Act of 2012 (enacted in January 2013) are not reflected in our analysis.

Similarly, the regulatory framework was assumed to take the regulations in place in 2011, and no regulatory measures beyond those adopted as of 2011 are imposed. The Clean Air Interstate Rule, for example, is in force for the purpose of our analysis, but other proposed regulations affecting power plants are not implemented. Our baseline assumption excluded important pending or recently finalized regulations such as the Cross-State Air Pollution Rule, the Mercury and Air Toxics Standards, and the proposed New Source Performance Standard for CO_2 on new power plants. Environmental Protection Agency vehicle emission standards in place as of 2011 are assumed to continue, meaning that the analysis does not include the impact of the vehicle standards adopted in August 2012. The Renewable Fuel Standards under the Energy Independence and Security Act are included, although these are modified to allow for a plausible potential waiver scenario.

To estimate the impacts of each of the specific tax provisions of interest, the selected models were run with a baseline assumption that all of the 2011 tax code provisions remained in place indefinitely, and then the models were rerun by removing each tax provision one at a time.

RESULTS OF THE COMMITTEE'S MODELING STUDIES

At the outset, it is useful to put the results described in this section in the context of projected GHG emissions under business-as-usual conditions and the magnitude of GHG reductions in certain proposed targets. For this purpose, we consider the difference between the emissions trajectory in our baseline and an "emissions reduction target" necessary to meet the climate-change targets proposed by the Obama administration in 2009 or analyzed in a recent comprehensive report by the National Academies in 2010. For the baseline emissions, we rely on the AEO 2011 estimates. The AEO 2011 baseline used for this report projects that CO_2 emissions will grow by 8 percent from 2010 to 6,105 MMT in 2035. This baseline reflects trends that are projected to occur without specific carbon-pricing policies such as cap-and-trade restrictions or taxes on GHG emissions. (We note as well that the AEO 2011 baseline has a lower emissions trajectory than many other integrated assessment models.) To meet the emissions reduction target would require that emissions be reduced 42 percent below 2005 levels by 2030. This is estimated to be a 60 percent reduction below the AEO11 baseline.

Energy-related Tax Expenditures

The committee identified 10 major energy tax expenditure provisions as candidates for analysis. Of these, the committee evaluated 5, which account for 71 percent of the revenue loss from the 10 provisions. This section provides a summary of four of the provisions, while the fifth (biofuels subsidies) is covered in subsequent sections.

Production Tax Credits for Renewable Electricity

The production and investment tax credits for renewable electricity provide a tax credit of 2.3 cents per kWh of power for the first 10 years of electricity production generated from qualifying renewable sources (primarily solar, wind, and biomass) or a credit equal to 30 percent of investment in qualifying equipment. These credits lower the cost of electricity generated from renewable resources, encouraging their substitution for fossil fuels and thereby tend to reduce GHG emissions.

The committee's analysis using the NEMS-NAS model indicates that these provisions lower CO_2 emissions. This finding holds for both the Reference and High-Macroeconomic-Growth scenarios. This finding confirms that decreasing the costs of low-carbon renewable fuels will lead to substitution away from high-carbon fossil fuels in the electricity sector. However, the impact is small, amounting to a reduction of 0.3 percent of U.S. CO_2 emissions compared to the reference scenario. Moreover, these tax expenditures are among the most costly that the committee examined in terms of revenue forgone per ton of CO_2 reduced.

Oil and Gas Depletion Allowances

The depletion allowance permits owners of oil and gas wells to deduct a value equal to the decline in the value of their reserves as oil or gas is extracted and sold—a method known as cost depletion. Under current law, some taxpayers may use percentage depletion as an alternative to cost depletion. Under percentage depletion, taxpayers deduct a percentage of gross income associated with the sale of the produced commodity. This deduction can exceed the cost of the original investment over the life of the property. Percentage depletion for oil and gas is currently available only for domestic production by independent (i.e., nonintegrated) companies up to a maximum of 1,000 barrels of oil per day (6 MMcf per day of natural gas) and cannot exceed half the net income from the property. The depletion rate is set at 15 percent of gross revenues associated with production.

In our modeling, removing the percentage depletion allowances (and substituting the lower valued with cost depletion) has virtually no effect on GHG emissions. The primary impact of percentage depletion on emissions comes from an increase in the production of natural gas and the associated ripple effects of higher natural gas production in other markets. Surprisingly, even though the depletion allowance is generally associated with oil, the NEMS-NAS model projects that it has virtually no impact on oil production.

Although natural gas production goes down as the tax preference is removed, the complex substitution patterns lead to largely offsetting forces and to a minimal impact on overall emissions. The average effect on GHG emissions over the time horizon of the models is too small to accurately estimate, or even determine, if the sign of the change is positive or negative.

Home Energy-efficient Improvement Credits

The committee examined qualitatively two other provisions. The first is Credits for Energy Efficiency Improvements to Existing Homes. Analysis of this provision proved difficult. The committee did not find, and was unable within its time and budget constraints to produce, detailed and reliable estimates of its impacts on GHG emissions. We expect that as the provision is currently structured, it is unlikely to produce major savings in GHG emissions. However, given the magnitude of the tax expenditure and the evidence of unexploited savings in this sector, we believe that understanding the impacts of tax incentives on household energy consumption should be a high priority for future research.

Nuclear Decommissioning Tax Preference

Another provision that was analyzed qualitatively was the special tax rate on reserves set up to decommission nuclear power plants at the end of their lifetime. The committee could find no detailed published studies of the impact of this provision on GHG emissions. Based on the available evidence, including the projections of nuclear power under different scenarios using the NEMS-NAS model, we find that the decommissioning provision is likely to have little impact on greenhouse gas emissions.

The underlying reasoning is straightforward. The capital costs for a new power reactor are high, about $7 billion in the most recent estimates (AEO

2012). Moreover, there have been only two reactors commissioned since the Three Mile Island accident in 1979. Projections used for this study found that there would be few or no new nuclear power plants built in the study period even with this provision in place. If the provision were removed, lowering the profitability of new nuclear plants, there would still be few or no plants built. Other factors might enter to change this conclusion, such as operating costs, lifetimes, and the expected price of natural gas for competing base-load power. But on the whole, the nuclear decommissioning tax preference provision is not considered significant enough to influence the decision by a utility company to make such a large capital expenditure.

Excise Taxes

In revenue terms, federal excise taxes are small compared to total tax expenditures. For fiscal year 2011, for example, total federal excise tax collections were $72 billion, while total tax expenditures amounted to $1,226 billion. However, most of the federal excise taxes are energy related, while only a small fraction of tax expenditures are energy related. Chapter 4 considers two such excise taxes—highway motor fuel taxes and taxes on air travel.

Highway Motor Fuel Taxes

The Internal Revenue Code levies highway fuels taxes of $0.184 per gallon of gasoline or alcohol fuel for on-road use, whether pure or blended. The Code also levies on a tax on diesel fuels at $0.244 per gallon for diesel and kerosene and $0.197 per gallon for diesel-water fuel emulsion.

This chapter reviewed four commissioned studies of the effect of removing the excise taxes on highway fuels. All four models find that removing the excise taxes on highway fuels would result in increasing greenhouse gas emissions. But the magnitude of the estimated effects varies dramatically for the different models.

Having studied the model results and the broader literature, the committee concludes that the differences among the models are large and incompletely understood. The differences arise from the types and values of price elasticities used by the different models, from assumptions about increasing biofuels production and consumption to meet the RFS mandates, from the volumetric bias of highway fuels taxes, and from application of the tax within each model's structure. A close examination of the results leads the committee to conclude that the NEMS-NAS and the FAPRI models capture the forces at work in this sector most reliably and therefore form the basis of our estimates. Taking these two modeling results together produces a striking conclusion: The impact of removing highway fuels taxes on GHG emissions is estimated to be very small because of special features of the taxes and the market. The volumetric bias of the taxes means that removing them favors ethanol, which will reduce the GHG impacts of increasing highway fuel consumption. Additionally, the renewable fuel standards constrain the use of ethanol. According to the two models, the effect of removing the highway fuels taxes is 4 MMT per year (NEMS) and 10 MMT per year (FAPRI). These are 0.07 percent and 0.17 percent of annual U.S. CO_2 emissions, respectively.

The committee emphasizes the contingent nature of the model projections. They are contingent because the results depend upon the structure, timing, and implementation of the renewable fuels standards (RFS) as well as a quirk in the tax structure (its volumetric bias). If the RFS were to disappear tomorrow, or if the regulations on E85 were to change drastically, or the volumetric bias of highway taxes were to be removed, the projected impacts of removing the gasoline tax might be substantially different and would probably be significantly larger. The magnitude of the differences across models leads the committee to caution against using the precise numerical results from a single model and recommends drawing only broad conclusions about the nature and direction of impacts. Policy makers and analysts should rely on multiple models, methodologies, and estimates in calculating impact of the tax code and other policies on greenhouse-gas emissions and climate change.

Aviation Fuel Taxes

The jet fuel excise tax is $0.043 per gallon for commercial aviation and $0.193 per gallon for noncommercial aviation. Even though airline excise tax rates have been raised in recent years, little research has been undertaken into their impacts, and particularly their impacts on GHG emissions. Only one of the modelers we commissioned, CBER, estimated the impact of removing the tax on jet fuel.

We believe that more work is needed in analyzing the economic impact and structure of aviation taxation. This is a sector producing rapidly growing GHG emissions, and as yet there are no ready substitutes for fossil-fuel-based jet fuel. Although the GHG emissions from aviation fuels are smaller than for motor fuels today, our results suggest that the GHG reductions per dollar of tax revenue foregone might be higher in this sector than for motor fuels.

Biofuels Provisions

One particularly important set of tax provisions involves the use of ethanol and other biofuels, particularly as substitutes for petroleum products. These provisions involve a complex combination of taxes, tax expenditures, import tariffs, and regulatory mandates that interact to change the composition of fuels.

Prior to 2012, the Internal Revenue Code provided an array of tax credits for biofuels. The most important was that alcohol fuels blended with gasoline or used in pure form as a fuel both qualified for a $0.45 per gallon credit under the Volumetric Ethanol Excise Tax Credit (VEETC). There were also tax credits for ethanol from small producers, for producers of cellulosic biofuels, for biodiesel, and for small agri-biodiesel producers. In addition, there was a $0.54 per gallon tariff on imported ethanol. These subsidies were the largest among all the energy-related ones. The VEETC and biodiesel provisions expired at the end of 2011, but under the committee's methodology, each of these provisions is included in our base case.

The excise tax exemption and credits lowered the cost of biofuels and therefore should have encouraged their substitution for petroleum motor fuels, reducing GHG emissions. Because biofuels are almost always sold as a blend with petroleum fuels, however, the subsidies also lowered the delivered price of the petroleum-biofuel blend, thereby encouraging additional consumption of motor fuels. Given the two factors operating in opposite directions and the fact that ethanol production has positive GHG emissions, the overall impact of the subsidy on GHG emissions is ambiguous.

The committee analyzed the biofuels provisions with two different models, although it concentrated its analysis on the FAPRI-MU model, which had the most detailed treatment of the biofuels sector. The findings indicate that removing all tax code provisions and the import tariff would result in a decrease of emissions of 5 MMT per year of CO_2 equivalent globally. This is less than 0.02 percent of global emissions. On a global basis, removal of the provisions results in a decrease in global ethanol use and an increase in global gasoline use. The results are complicated by the mandates for renewable fuels. If the mandates are removed along with the subsidies, the estimated emissions are smaller than the estimates with the mandates. Therefore, as is intuitive, the mandates reduce (in absolute value) the size of the impact of the subsidy on GHGs. The results of the other modeling studies are consistent with the central FAPRI estimates.

These results show the often-counterintuitive nature of the effects of tax subsidies. Although it might seem obvious that subsidizing biofuels should reduce CO_2 emissions because they rely on renewable resources rather than fossil fuels, many studies we reviewed found the opposite. As structured, the biofuels tax credits encouraged the consumption of motor fuels overall because they lower prices, and this effect appears to offset any reduction in the GHG intensity of motor fuels when switching from gasoline to biofuels. (The GHG intensity is calculated as the ratio of the emissions of GHG per unit output.)

Broad-based Tax Expenditures

The committee examined a limited set of broad-based tax expenditures in addition to energy-sector provisions. The broad-based provisions are important primarily because of their impact on the size and composition of the overall economy. First, they may shift output from industries with lower GHG intensities to industries with higher GHG intensities or vice versa. Second, they may affect the rate of growth of national output and therefore change emissions simply because the economy is larger or smaller.

Although there are many large tax expenditures, here as elsewhere the committee had to narrow the scope of its inquiry. After examining the largest tax expenditures (see Table 1-3), the committee decided to examine three broad types of provisions: (1) tax incentives that affect investment generally, (2) tax incentives for owner-occupied housing, and (3) tax incentives for provisions of health insurance. These revenue losses from these provisions totaled $370 billion in fiscal year 2012, or about one-third of all tax expenditures.

We undertook our analysis using IGEM, which is a full-employment, multisector, general-equilibrium model. The large revenue increases entailed in removing these provisions were offset or recycled in two different ways—by lump-sum transfers and by changes in individual and corporate tax rates. Because the model has large incentive effects of changes in the tax structure, IGEM suggested that some of the largest estimated impacts of changes in the tax provisions on GHG emissions come through changes in overall economic activity.

Accelerated Depreciation

Accelerated depreciation is one of the largest business tax expenditures in the federal income tax code. This set of provisions allows businesses to write off the value of their capital assets at a rate that is faster than the estimated economic depreciation. This provision is the only one of the broad-based provisions for which the committee has confidence that the model is able to capture the principal channels through which GHG emissions are affected. The model runs show that eliminating accelerated depreciation would reduce the GHG intensity of national output by shifting production away from GHG-intensive activities such as coal mining and electric power generation to low-GHG activities such as communications.

However, the net effect depends upon how the resulting revenues are recycled. Eliminating accelerated depreciation would reduce the capital stock and national output. The effect on the capital stock would be partially offset if labor supply increases. Furthermore, if the additional revenue were used to finance cuts in individual and corporate marginal tax rates, this might eliminate any net impact on GHGs. If, on the other hand, the higher revenues are returned to taxpayers in lump-sum rebates, the analyses show that overall emissions are about 2 percent lower, reflecting the combination of lower emissions per dollar of national output and lower national output.

Owner-occupied Housing Provisions

The federal income tax provides significant tax incentives for investments in owner-occupied housing. These include deductibility of mortgage interest and property taxes and exclusion from taxation of the first $250,000 ($500,000 for couples) of capital gains on home sales. Defenders of these preferences often justify them as encouraging more people to own homes, although some economic studies indicate that the major effect is on home size rather than housing tenure choice. These are large tax preferences, totaling $147 billion in 2011, according to the Treasury.

According to the estimates prepared for the committee, eliminating the tax subsidies for owner-occupied housing and using the revenue to lower marginal tax rates would improve the efficiency of allocation of the capital stock and increase national output. GHG emissions would increase at about the same rate as GDP increases. These results are consistent with earlier analyses.

However, the simulation does not fully capture the effects of the housing subsidies on emissions intensity. In particular, it does not capture the effects on housing size and in turn household consumption of electricity, natural gas, and home heating oil. It also is unable to capture potential effects of the housing tax preferences on the spatial distribution of housing and patterns of automobile use and gasoline consumption. Even the relationship between housing subsidies and changes in the composition of output are somewhat obscure in the IGEM model. We therefore find the analysis of the housing provisions inconclusive and believe that the effects on GHG emissions of eliminating or reducing the housing tax preferences are a high-priority topic for further research.

Employer-provided Health Care Provisions

The largest single tax expenditure in the Internal Revenue Code is the exclusion of employer-provided health insurance (EPHI) benefits from taxable income of employees. The Treasury Department estimates that exclusion of EPHI benefits will reduce income tax receipts by $180 billion in 2012.

The only model considered by the committee that had the potential to estimate the impact of health care provisions was IGEM. However, this model does not capture the pathways by which the exclusion of EPHI from the tax on employee compensation might affect demand for health care services. Therefore, simulations for the committee using IGEM treated the removal of the health tax preference as equivalent to imposing a new excise tax on personal and business services (of which health is a sector). The model found that eliminating the EPHI exemption would raise national output if offset by cuts in marginal tax rates, but would reduce national output slightly if revenues are rebated in a lump-sum fashion.

The model's findings of the effect of changes in the composition of output on GHG intensity were anomalous. We expected that eliminating health care subsidies would raise GHG emissions per unit of output because the health care sector is less GHG intensive than the rest of the economy. The IGEM model results show the opposite effect, however, with a small decrease in GHG intensity. Our inability to understand the structural features of the model that explain these results lead us to conclude that the impact of the health provisions on GHG emissions remains an open question and an important subject for future research.

Further Observations on the Broad-based Tax Expenditures

The committee's major finding is that the broad-based provisions influence GHG emissions primarily through their effects on overall national output. In most cases, the change in GHG emissions was close to or equal to the change in national output induced by removing the tax provision (see particularly Table 6-2). This is a plausible conclusion following from the relatively modest effects of all the tax provisions on the ratio of emissions to national output. If tax law changes do not affect the emissions-output ratio markedly, then their effect on output will be the principal driver of their impact on emissions.

A second finding is that the way revenues generated by eliminating tax preferences are recycled significantly affects output and emissions. Recycling new revenues through reductions in individual and company tax rates is likely to raise national output and therefore will also increase GHG emissions. Estimates from the IGEM model indicate that removing the broad-based tax expenditures might increase national output and related GHG emissions in the order of 1 to 2 percent of baseline emissions. In contrast, lump-sum tax rebates do not increase incentives to save, invest, work more, or use resources more efficiently and will have little effect on national output. This result also emphasizes the importance of examining the sources of the revenue when analyzing the impact of different tax expenditures and subsidies.

A third finding is that the broad-based provisions generally have little effect on emissions intensities, ranging from -0.3 percent to +0.2 percent over the period 2010-2035. Thus, whatever the uncertainty of our results, the one common finding is that the impacts of these provisions on emissions are not large.

Finally, we need to reiterate our reservations about the simulations of the broad-based tax expenditures. The results are highly sensitive to assumptions about how tax revenues from eliminating the provisions are returned to the economy. Moreover, we were able to use only a single model to calculate the impacts of most of these provisions, and that model is not transparent about some of the pathways through which tax provisions affect GHG emissions changes. Additionally, in the case of accelerated depreciation, where we were able to compare IGEM's results with other modeling results, they were divergent.

We conclude that changes in broad-based tax provisions are likely to have a small impact on overall GHG emissions, mainly by changing national output. Even the largest impact on emissions, however, is tiny relative to the projected growth in GHG emissions over the period 2010–2035. In its baseline simulation, the IGEM model projects a 43 percent growth in emissions over that period. The largest change in emissions from any of the broad-based provisions was in the order of 2 percent. Therefore, the tax provision with the largest emissions impact changes the growth of emissions over the period by only a tiny fraction.

Comparison with CBER Modeling Results

We compared the results of our detailed modeling studies with those of a comprehensive study of energy tax expenditures by the CBER modeling group at the University of Nevada, Las Vegas. Their initial report was released at about the time our study was launched. The committee reviewed their work and commissioned additional simulations using their model.

As we observed in Chapter 2, the advantage of the CBER approach is its comprehensive nature. It is also a transparent modeling structure, in which the assumptions are laid out clearly in a way that allows users to understand key driving forces. It has the disadvantages of any partial equilibrium model; in addition, it lacks many of the important regulatory and fine-grained structures of the more detailed energy-sector models. The study also includes some non-tax direct expenditures, such as spending on low-cost residential weatherization. It studies the energy tax expenditures in effect for the period 2005–2009, which is different from the other models. On the other hand, the average annual cost of those expenditures was $19 billion, approximately the same as those in our study.

The committee used the CBER model to obtain an order-of-magnitude estimate of the impact of all energy-related subsidies. Under the methods and assumptions of that study, if all tax subsidies would have been removed, then net CO_2 emissions would have decreased by 30 MMT per year over the 2005-2009 period. This total represented about ½ percent of total U.S. CO_2 emissions over this period.

The CBER study found that several of the tax policies reduced CO_2 emissions, but that direct expenditures reduced emissions more than tax subsidies. Consistent with our modeling results, CBER found the biofuels credits increased CO_2 emissions.

The committee concludes that the basic finding of the CBER study that removing all energy-sector subsidies would increase GHG emissions is plausible, but it is not robust and is subject to a large margin of error. The CBER results are consistent with the basic findings of the detailed modeling studies we conducted—that the overall effect of current energy tax subsidies is close to zero.

We summarize the results of the commissioned economic modeling simulations in Table 7-1.

SUMMARY OF FINDINGS

The committee's report on the impacts of the U.S. tax code on greenhouse gas emissions has relied on analyses that ranged from simplified modeling approaches through detailed energy-sector modeling to general equilibrium economy-wide modeling. The estimates have major uncertainties for reasons that we have described at various points in this report. Nevertheless, despite these uncertainties, we draw the following general conclusions.

TABLE 7-1 Summary of Modeling Results

Model Parameters	NEMS	CBER	FAPRI	IGEM
Baseline Parameters and Assumptions	AEO 2011[1]	AEO 2012[2]	AEO 2012	AEO 2011 for GDP and emissions baseline Other variables as specified in Chapter 6
Gases Included in Model	Only CO_2	Only CO_2	CO_2, CH_4, N_2O, LUC/ILUC	CO_2, CH_4, N_2O, HGWPs
Policies Modeled and Indicative Results				
Model Policy	NEMS	CBER	FAPRI	IGEM
Tax Credits for Production of Renewable Electricity	Decrease	Decrease	**Not Modeled**	**Not Modeled**
Excess of Percentage Over Cost Depletion	Increase	Increase	**Not Modeled**	**Not Modeled**
Credits for Energy Efficiency Improvements to Existing Homes	**Not Modeled**	Decrease	**Not Modeled**	**Not Modeled**
Special Tax Rate on Nuclear Decommissioning Reserve Funds	**Not Modeled**	Decrease	**Not Modeled**	**Not Modeled**

[1] AEO 2011 Assumptions: GDP 2.7%; GDP components of final demand – Consumption 2.4%; Investment 4.6%; Government 0.7%; Exports 6.3%; Imports 4.6%.
[2] AEO 2012 Assumptions: GDP 2.6%; GDP components of final demand – Consumption 2.3%; Investment 4.2%; Government 0.4%; Exports 6.0%; Imports 4.1%.

Highway Motor Fuels Excise Tax	Decrease	Decrease	Decrease	Decrease
Volumetric Ethanol Excise Tax Credit	Very small impact	Increase	Increase	**Not Modeled**
Ethanol-specific Tariff	Very small impact	**Not Modeled**	Increase	**Not Modeled**
Biodiesel Excise Tax Credit	Very small impact	Decrease	Increase	**Not Modeled**
Tax Incentives for Home Ownership	**Not Modeled**	**Not Modeled**	**Not Modeled**	Uncertain
Tax Incentives for Health Insurance and Health Care	**Not Modeled**	**Not Modeled**	**Not Modeled**	Uncertain
Accelerated Depreciation	Decrease	**Not Modeled**	**Not Modeled**	Increase

First, the combined effect of current energy-sector tax expenditures on GHG emissions is very small and could be negative or positive. The most comprehensive study available suggests that their combined impact is less than 1 percent of total U.S. emissions. If we consider the estimates of the effects of the provisions we analyzed using more robust models, they are in the same range. We cannot say with confidence whether the overall effect of energy-sector tax expenditures is to reduce or increase GHG emissions. (Chapters 3-5)

Second, individual energy-sector tax expenditures in some cases contribute to, and in other cases subtract from, U.S. or global GHG emissions. The subsidies on ethanol that expired at the end of 2011 are estimated to increase global GHG emissions. By contrast, the production and investment tax credits for renewable electricity appear to reduce U.S. GHG emissions. The depletion allowance has virtually zero impact on emissions. (Chapters 3-5)

Third, the best existing analytical tools are unable to determine in a reliable fashion the impact of some important subsidies. Important tax expenditures that have resisted analysis include ones subsidizing residential energy efficiency. The difficulties in this case involve such factors as the discount rate consumers apply to future fuel savings, the strength of any rebound effect, and the extent to which consumers understand and respond to tax law changes. (Chapters 3-5)

Fourth, the revenues foregone by energy-sector tax subsidies are substantial in relation to the effects on GHG emissions. The Treasury estimates that the revenue loss from energy-sector tax expenditures in fiscal years 2011 and 2012 totaled $48 billion. Few of these were enacted to reduce GHG emissions. To the extent that they include as a primary objective reducing GHG emissions, however, they are inefficient. Very little if any GHG reductions are achieved at substantial cost with these provisions. (Chapters 1 and 3-5)

Fifth, the impacts of the broad-based tax expenditures on GHG emissions come primarily through their impact on the level of national output. Broad-based tax expenditures entail roughly 50 times more revenues foregone than the energy-sector subsidies. We investigated a subset of provisions representing about one-third of the revenue losses from tax expenditures—subsidies to equipment investment through accelerated depreciation, to health care, and to owner-occupied housing. Except for accelerated depreciation, we were unable to reach a definite conclusion on whether they increase or decrease GHG emissions per unit of output. Rather, the principal effect of these provisions is on the size of national output and the allocation of resources among sectors. If the revenue impacts of removing broad-based subsidies were offset by reducing distortionary taxes, the resulting increase in national output would be accompanied by increased GHG emissions. If the subsidies were replaced with lump-sum tax cuts that do not reduce distortions, there would likely be little effect on national output or emissions. (Chapter 6)

Sixth, it is difficult to estimate the impact of the broad-based tax expenditures on GHG emissions intensity. The committee examined the existing literature and commissioned modeling studies to estimate the effects of changes in the broad-based provisions on the overall GHG intensity of the economy. The results were not judged to be sufficiently reliable to draw firm conclusions. (Chapter 6)

Seventh, the effects of many tax provisions are complicated by their interaction with regulations. Very few tax provisions take place in a regulatory vacuum. Particularly in the energy sector, energy and environmental regulations overlay and interact with tax provisions. Prime examples are the interaction of tax provisions with the CAFE standards for light-duty vehicles, the air pollution standards for the mix of electricity generation, the Renewable Portfolio Standards for electricity generation, and the Renewable Fuel Standards for motor fuels blended from petroleum and ethanol. An important finding of our studies is that regulatory and environmental constraints generally reduce the size of the impacts of tax provisions on GHG emissions. (Chapters 3-5)

Eighth, energy excise taxes reduce GHG emissions, but the impact is limited because of special features of the tax and because of regulatory constraints. Analysts have studied the effects of gasoline taxes for many years. The committee's estimates show unambiguously that highway fuel excise taxes reduce fuel consumption and GHG emissions. The analysis for this report finds that the current highway fuels taxes have a relatively small impact on GHG emissions because of the volumetric bias of the taxes as well as the constraints imposed by the renewable fuels standards. (Chapter 4)

RESEARCH RECOMMENDATIONS

After reviewing various approaches, the committee concludes that the only reliable methodology for estimating the impacts of tax policy on GHG emissions is through the use of energy-economic modeling. However, existing models are primarily useful for understanding the economic and energy-sector linkages that produce emissions changes and for suggesting the likely sign of those effects. In very few cases are existing models able to determine with precision the quantitative impacts of tax provisions on GHGs. Moreover, there are numerous shortcomings to existing models. The following recommendations to the Congress, the modeling community, the research support agencies, as well as the broader community provide guidance on the areas where the committee finds that more attention is needed.

The committee recommends continued support of energy-economic modeling to better understand the impacts of taxes and other public policies on greenhouse gas emissions and the broader economy. Particular attention should be given to improving current models in the following ways:

First, models need to be made more transparent by clarifying both their assumptions and their structure.

Second, models should include measures of economic welfare that can be used to measure the efficiency and distributional impacts of policies.

Third, there should be more work to integrate partial equilibrium models with general equilibrium models so that the impact of revenue recycling and overall economic impacts can be more reliably estimated.

At present, the impacts of taxes and tax expenditures on GHG emissions are difficult to measure reliably. This proved to be the case across the entire spectrum of provisions evaluated in this study. The detailed analyses in Chapters 3 through 6 show the difficulties the committee encountered in arriving at reliable estimates of GHG impacts, and in some cases the committee was unable to determine whether the sign of the impact was negative or positive. The major way to determine the reliability of large energy-economic models today is through comparing different approaches. This leads to a final research recommendation:

Fourth, the committee recommends increased attention to studies that compare energy-economic models as a tool for improving understanding of models, narrowing the range of estimates, and improving model reliability.

GUIDANCE FOR SCORING GHG EMISSIONS

The Congressional Budget Office is required to estimate ("score") the budget impact of proposed legislation. Some have suggested that there be a parallel procedure for estimating the effects of legislation on GHG emissions and climate change. The committee finds that attempting to institute scoring is premature. Estimating the impacts of tax provisions on greenhouse gas emissions is difficult because the mechanisms are so complex; there are so many interacting forces; the regulatory and market environments change quickly and unpredictably; and regulatory and tax arbitrage across national boundaries tends to reduce or increase the impact measured from a global vantage point. Today's modeling capabilities are not yet up to this task, particularly on the rapid schedule necessary for legislative action. Simply put, we do not believe that GHG accounting of taxes and expenditures is ready for prime time; indeed, it is not even ready for a short cameo spot on "The Accounting Channel" in the wee hours of the night.

Because of the difficulties and resources required to provide reliable estimates, the committee discourages requiring the formal scoring of tax proposals for their impacts on GHG emissions. Much further work needs to be done before it can be accomplished routinely and reliably.

USE OF TAX POLICY TO ADDRESS CLIMATE CHANGE

In addition to estimating the impacts of the tax code on GHG emissions, the committee was asked to examine broader implications of taxes and climate-change policy. The congressional Joint Committee on Taxation, in its report on the legislation authorizing this study, suggested that the National Academies' report "discuss the importance of controlling carbon dioxide and greenhouse gas emissions as part of a comprehensive national strategy for reducing U.S. contributions to global climate change, and evaluate the potential for changes in the Code to reduce carbon dioxide emissions" (Joint Committee on Taxation, 2009). Although the committee does not make any recommendations about specific changes, the analysis undertaken for this report leads to several important insights and cautions about tax policy in the context of climate change.

First, current tax expenditures and subsidies are a poor tool for reducing greenhouse gases and achieving climate-change objectives. The committee has found that several existing provisions have perverse effects, while others yield little reduction in GHG emissions per dollar of revenue loss. This is not surprising, because most of the provisions act indirectly rather than directly on emissions and most provisions were not structured with reduction of GHG emissions as a primary policy objective. The feedback effects within the energy sector (e.g., the fuel substitution effects when tax policy favors one source over others) or the international spillover effects (e.g., shifts in trade flows due to differential tax treatment) can offset or even reverse the expected direct effects of these policies. Such leakages and regulatory and tax arbitrage are common features of indirectly targeted provisions. For example, it may seem obvious that a subsidy for renewable fuels will reduce GHG emissions, but the response of different sectors can undermine the goal, as our analysis of biofuels shows. Thus, if tax expenditures are to be made an effective tool for reducing GHG emissions, much more care will need to be applied to designing the provisions to avoid inefficiencies and perverse offsetting effects. (Chapters 2-6)

Second, some tax expenditures are more efficient than others. Some tax policies have fewer leakages and offsetting forces than others and therefore may be more effective. In contrast with the ethanol subsidies, where the emissions from biofuels more than offset the reduction in petroleum emissions, the renewable production tax credits for electricity appear to have the desired effects of increasing the production of renewables in the mix of energy sources and reducing GHG emissions. At their current scale, however, they achieve small emission reductions and are costly per unit of emissions reduction. (Chapters 3-5)

Third, the committee's reservations about tax expenditures and subsidies do not necessarily apply to tax incentives directly targeted on activities such as research and development on technological advances. Governmental support for innovation will help the nation and the world transition to a low-carbon energy system. We did not review these expenditures nor explore the effectiveness of the current incentives to promote low-carbon technologies, or the best way to deliver them, but there is a substantial literature justifying R&D subsidies on the basis of its positive spillovers and private underinvestment in research in the energy sector. (Chapter 2)

Fourth, tax reforms that increase the economic efficiency of our economy may increase GHG emissions, but the increased output is likely much more than sufficient to pay for reducing the higher emissions if efficient climate-change policies are employed to reduce emissions. Our general equilibrium modeling suggests that substituting a less distortionary tax regime for current tax expenditures may well raise GHG emissions as the economy grows faster. This is not an argument against removing inefficient tax expenditures, however. Effective tax reform will increase the nation's productivity and living standards, thereby providing more than sufficient resources to pay for reducing the additional GHG emissions. (Chapter 2, EIA, 2009; National Research Council, 2010; Clarke et al., 2010)

Finally, many studies have found that the most reliable and efficient way to achieve given climate-change objectives is to use direct tax or regulatory policies that create a market price for CO$_2$ and other greenhouse gas emissions. A central finding of many studies in this area is that the most efficient way to reduce GHG emissions is through policies that create a market price for CO$_2$ and other GHGs (National Research Council, 2010; Resources for the Future, 2010). This can be accomplished either by tradable GHG-emissions allowances or by taxes on GHG emissions. The national and global emissions reductions necessary to meet internationally agreed-upon climate objectives are many times larger than those resulting from current tax subsidies. The committee finds that tax policy can make a substantial contribution to meeting the nation's climate-change objectives, but that the current approaches will not accomplish that. In order to meet ambitious climate-change objectives, a different approach that targets GHG emissions directly through taxes or tradable allowances will be both necessary and more efficient. (Chapter 2, EIA, 2009; National Research Council, 2010; Clarke et al., 2010)

References

Adams, D.M., R.J. Alig, J.M. Callaway, B.A. McCarl, and S.M. Winnett. 1996. The Forest and Agricultural Sector Optimization Model (FASOM): Model Structure and Policy Applications. Research Paper PNW-RP-495. U.S. Department of Agriculture, Forest Service, Pacific Northwest Research Station, Portland, OR [online]. Available: http://www.fs.fed.us/pnw/pubs/pnw_rp495.pdf [accessed June 5, 2013].

Aghion, P., and P. Howitt. 1998. Endogenous Growth Theory. Cambridge, MA: MIT Press.

Airlines for America. 2012. Taxes and Fees [online]. Available: http://www.airlines.org/Pages/Taxes-and-Fees.aspx [accessed September 17, 2012].

Allaire, M., and S. Brown. 2011. U.S. Energy Subsidies: Effects on Energy Markets and Carbon Dioxide Emissions, July 11, 2011 [online]. Available: http://emf.stanford.edu/files/docs/323/StephenBrown-Paper.pdf [accessed June 5, 2013].

Arrow, K.J. 1962. Economic welfare and the allocation of resources for invention. Pp. 609-626 in The Rate and Direction of Inventive Activity, R. Nelson, ed. National Bureau of Economic Research [online]. Available: http://www.nber.org/chapters/c2144.pdf [accessed June 5, 2013].

Babcock, B.A., and M. Carriquiry. 2010. An Exploration of Certain Aspects of CARB's Approach to Modeling Indirect Land Use from Expanded Biodiesel Production. Staff Report 10-SR 105. Iowa State University, Center for Agricultural and Rural Development, Ames, IA.

Baumol, W.J., and W.E. Oates. 1988. The Theory of Environmental Policy, 2nd Ed. Cambridge, UK: Cambridge University Press.

Beach, R., and B. McCarl. 2010. U.S. Agricultural and Forestry Impacts of the Energy Independence and Security Act: FASOM Results and Model Description. Final Report. Prepared for Office of Transportation and Air Quality, U.S. Environmental Protection Agency, Washington, DC, by RTI International, Research Triangle Park, NC.

Beach, R., D. Adams, R. Alig, J. Baker, G.S. Latta, B. McCarl. B.C. Murray, S.K. Rose, and E. White. 2013. Model Documentation for the Forest and Agricultural Sector Optimization Model with Greenhouse Gases (FASOMGHG). U.S. Environmental Protection Agency, Washington, DC.

Bento, A.M., L.H. Goulder, M.R. Jacobsen, and R.H. von Haefen. 2009. Distributional and efficiency impacts of increased U.S. gasoline taxes. Am. Econ. Rev. 99(3):667-699.

Berstein, P.M., R.L. Earle, and W.D. Montgomery. 2007. The role of expectations in modeling costs of climate change policies. Pp. 216-226 in Human-Induced Climate Change: An Interdisciplinary Assessment, M.E. Schlesinger, H.S. Kheshgi, J.B. Smith, F.C. de la Chesnaye, J.M. Reilly, T. Wilson, and C. Kolstad, eds. Cambridge, UK: Cambridge University Press.

Bogdanski, J.A. 2011. Reflections on the environmental impacts of federal tax subsidies for oil, gas, and timber production. Lewis and Clark Law Review (April 15):323-337 [online]. Available: http://www.lclark.edu/live/files/8324-lcb152art2bogdanski [accessed June 5, 2013].

Bohi, D.R. 1981. Analyzing Demand Behavior. Baltimore, MD: John Hopkins University Press.

Bohi, D.R., and M. Zimmerman. 1984. An update of econometric studies of energy demand. Annu. Rev. Energy 9:105-154.

Borenstein, S. 2011. Why can't U.S. airlines make money? Am. Econ. Rev. 101(3):233-237.

Bosetti, V., C. Carraro, M. Galeotti, E. Massetti, and M. Tavoni. 2006. WITCH: A World Induced Technical Change Hybrid Model. Pp. 13-38 in Hybrid Modeling of Energy Environment Policies: Reconciling Bottom-up and Top-down, a Special Issue of the Energy Journal. Cleveland, OH: Energy Economics Education Foundation, Inc.

Busse, M.R., C.R. Knittel, and F. Zettelmeyer. 2013. Are consumers myopic? Evidence from new and used car purchases. Am. Econ. Rev. 103(1):220-256.

Center for Global Development. 2007. Carbon Dioxide Emissions From Power Plants Rated Worldwide. Science Daily, November 15, 2007 [online]. Available: http://www.sciencedaily.com/releases/2007/11/071114163448.htm [accessed June 9, 2012].

Chakravorty, U., M.H. Hubert, and B.U. Marchand. 2012. Does the U.S. Biofuel Mandate Increase Poverty in India? Association of Resource and Environmental Economics 2nd Annual Summer Conference, June 3-5, 2012, Asheville, NC [online]. Available: http://www2.toulouse.inra.fr/lerna/seminaires/Papier_Chakravorty.pdf [accessed June 5, 2013].

Clarke, L., J. Edmonds, V. Krey, R. Richels, S. Rose, and M. Tavoni. 2009. International climate policy architectures: Overview of the EMF 22 International Scenarios. Energ. Econ. 31(suppl. 2):S64-S81.

Dahl, C.A. 1993. A survey of oil demand elasticities for developing countries. OPEC Rev. 17(4):399-420.

Dahl, C.A. 2002. Energy demand and supply elasticities. In Encyclopedia of Life Support Systems (EOLSS). Oxford, UK: EOLSS Publishers [online]. Available: http://www.eolss.net/Sample-Chapters/C08/E3-21-02-04.pdf [accessed June 5, 2013].

Dahl, C.A. 2012. Measuring global gasoline and diesel price and income elasticities. Energ. Policy 41(February):2-13.

Dahl, C.A., and T. Duggan. 1998. Survey of price elasticities from economic exploration models of U.S. oil and gas supply. J. Energy Financ. Dev. 3(2):129-169.

Debreu, G. 1959. Theory of Value: An Axiomatic Analysis of Economic Equilibrium. New Haven: Yale University Press.

Decreux, Y., and H. Valin. 2007. MIRAGE: Updated Version of the Model for Trade Policy Analysis: Focus on Agriculture and Dynamics. CEPII Working Paper N2007-15 [online]. Available: http://manoa.hawaii.edu/ctahr/aheed/ALex/AHEED_Ref_MIRAGE_Description.pdf [accessed June 5, 2013].

Devadoss, S., P. Westhoff, M. Helmar, E. Grundmeier, K. Skold, W. Meyers, and S.R. Johnson. 1993. The FAPRI modeling system at CARD: A documentation summary. Pp. 129-150 in Agricultural Sector Models for the United States, S.R. Taylor, K.H. Reichelderfer, and S.R. Johnson, eds. Ames, IA: Iowa State University Press.

EIA (U.S. Energy Information Administration). 2011. Electricity market module: Technological optimism and learning. P. 98 in Assumptions to the Annual Energy Outlook. U.S. Department of Energy, Washington, DC.

EIA (U.S. Energy Information Administration). 2011. Annual Energy Outlook 2011 with Projections to 2035. DOE/EIA-0383(2011). U.S. Energy Information Administration, Washington, DC [online]. Available: http://www.columbia.edu/cu/alliance/do cuments/EDF/Wednesday/Heal_material.pdf [accessed June 6, 2013].

EIA (U.S. Energy Information Administration). 2012. Annual Energy Outlook 2012 with Projections to 2035. DOE/EIA-0383(2012). U.S. Energy Information Administration, Washington, DC [online]. Available: http://www.eia.gov/forecasts/aeo/pdf/03 83(2012).pdf [accessed June 6, 2013].

EPA (U.S. Environmental Protection Agency). 2007. Greenhouse Gas Impacts of Expanding Renewable and Alternative Fuel Use. EPA-420-F-07-035. Office of Transportation and Air Quality, U.S. Environmental Protection Agency. April 2007.

EPA (U.S. Environmental Protection Agency). 2009. EPA Analysis of the American Clean Energy and Security Act of 2009, H.R. 2454 in the 111th Congress, June 23, 2009. Office of Atmospheric Programs, U.S. Environmental Protection Agency [online]. Available: http://www.epa.gov/climatechange/Downloads/EPAactivities/HR2454_A nalysis.pdf [accessed June 6, 2013].

EPA (U.S. Environmental Protection Agency). 2012. Inventory of U.S. Greenhouse Gas Emissions and Sinks: 1990-2010. EPA 430-R-12-01. U.S. Environmental Protection Agency, Washington, DC [online]. Available: http://www.epa.gov/climatecha nge/Downloads/ghgemissions/US-GHG-Inventory-2012-Main-Text.pdf [accessed June 6, 2013].

Espey, J.A., and M. Espey. 2004. Turning on the lights: A meta-analysis of residential electricity demand elasticities. J. Agr. Appl. Econ. 36(1):65-81.

Espey, M. 1998. Gasoline demand revisited: An international meta-analysis of elasticities. Energ. Econ. 20(3):273-295.

Gale, W., J. Gruber, and S. Stephens-Davidowitz. 2007. June 16. Encouraging Homeownership through the Tax Code. Tax Notes (June 18):1171-1189 [online]. Available: http://www.taxpolicycenter.org/UploadedPDF/1001084_Encouraging_Homeowners hip.pdf [accessed June 5, 2013].

GAO (U.S. Government Accountability Office). 2005. Tax Expenditures Represent a Substantial Federal Commitment and Need to Be Re-examined, September 23, 2005[online]. Available: http://www.gao.gov/products/GAO-05-690 [accessed June 6, 2013].

Gillingham, K.T. 2011. The Consumer Response to Gasoline Price Changes: Empirical Evidence and Policy Implications. Ph.D. Dissertation, Department of Management Science & Engineering, Stanford University [online]. Available: http://purl.stanfo rd.edu/wz808zn3318 [accessed June 5, 2013].

Goettle, R.J., M.S. Ho, D.W. Jorgenson, D.T. Slesnick, and P.J. Wilcoxen. 2007. IGEM, an Inter-temporal General Equilibrium Model of the U.S. Economy with Emphasis on Growth, Energy and the Environment [online]. Available: http://scholar.harv ard.edu/files/jorgenson/files/igem_documentation-1.pdf [accessed June 5, 2013].

Greening L.A., D.L. Greene, and C. Difiglio 2000. Energy efficiency and consumption – the rebound effect – a survey. Energ. Policy 28(6-7):389-401.

Greenstone, M., and H. Allcott. 2012. Is there an energy efficiency gap? J. Econ. Perspect. 26(1):3-28.

Gurgel, A., J. Reilly, and S. Paltsev. 2007. Potential land use implications of a global biofuels industry. J. Agr. Food Ind. Organ. 5(2):Art. 9.

Gurgel, A., T. Cronin, J. Reilly, S. Paltsev, D. Kicklighter, and J. Melillo. 2011. Food, fuel, forests and the pricing of ecosystem services. Am. J. Agr. Econ. 93(2):342-348.

Hertel, T. 2011. The global supply and demand for land in 2050: A perfect storm? Am. J. Agr. Econ. 93(2):259-275.

Hertel, T., W. Tyner, and D. Birur. 2010. The global impacts of biofuel mandates. Energy J. 31(1):75-100.

Hicks, J.R. 1932. The Theory of Wages. London: Macmillan.

Hirst, E., W. Fulkerson, R. Carlsmith, and T. Wilbanks 1982. Improving energy efficiency: The effectiveness of government action. Energ. Policy 10(2):131-142.

Hitaj, C. 2012. Wind Power Development in the United States. Working paper. Department of Agricultural and Resource Economics University of Maryland [online]. Available: http://terpconnect.umd.edu/~creitmai/Hitaj_WindPowerDevelopment.pdf [accessed June 6, 2013].

IPCC (Intergovernmental Panel on Climate Change). 1990. Report prepared for Intergovernmental Panel on Climate Change by Working Group I (1990). J.T. Houghton, G.J. Jenkins and J.J. Ephraums (eds.). Cambridge University Press

IPCC (Intergovernmental Panel on Climate Change). 1996. Climate Change 1995: The Science of Climate Change. J.T. Houghton, L.G. Meira Filho, B.A. Callander, N. Harris, A. Kattenberg and K. Maskell eds. Contribution of Working Group I to the Second Assessment Report of the Intergovernmental Panel on Climate Change. Cambridge University Press.

IPCC (Intergovernmental Panel on Climate Change). 2002. Climate Change 2001: The Scientific Basis. J.T. Houghton, Y. Ding, D.J. Griggs, M. Noguer, P.J. van der Linden, X. Dai, K. Maskell, C.A. Johnson. Contribution of Working Group I to the Third Assessment Report of the Intergovernmental Panel on Climate Change. Cambridge University Press .

IPCC (Intergovernmental Panel on Climate Change). 2001. Climate Change 2001: Mitigation: A Report of Working Group III of the Intergovernmental Panel on Climate Change. Cambridge, UK: Cambridge University Press [online]. Available: http://www.grida.no/publications/other/ipcc_tar/?src=/climate/ipcc_tar/wg3/index.htm [accessed June 6, 2013].

IPCC (Intergovernmental Panel on Climate Change). 2007. Climate Change 2007: The Physical Science Basis, Contribution of Working Group I to the Fourth Assessment Report of the Intergovernmental Panel on Climate Change, S. Solomon, D. Qin, M. Manning, Z. Chen, M. Marquis, K.B. Averyt, M. Tignor, and H.L. Miller, eds. Cambridge, UK: Cambridge University Press [online]. Available: http://www.ipcc.ch/publications_and_data/publications_ipcc_fourth_assessment_report_wg1_report_the_physical_science_basis.htm [accessed June 6, 2013].

IPCC (Intergovernmental Panel on Climate Change). 2012. Managing the Risks of Extreme Events and Disasters to Advance Climate Change Adaptation: Special Report of the Intergovernmental Panel on Climate Change, C.B. Field, V. Barros, T.F. Stocker, Q. Dahe, D.J. Dokken, K.L. Ebi, M.D. Mastrandrea, K.J. Mach, G.K. Plattner, S.K. Allen, M. Tignor, and P.M. Midgley, eds. Cambridge, UK: Cambridge University Press [online]. Available: http://ipcc-wg2.gov/SREX/images/uploads/SREX-All_FINAL.pdf [accessed June 6, 2013].

IRS (U.S. Internal Revenue Service). 2012. Table 20. Federal Excise Taxes Reported to or Collected by the Internal Revenue Service, Alcohol and Tobacco Tax and Trade Bureau, and Customs Service, by Type of Excise Tax, Fiscal Years 1999-2010 [online] Available: http://www.irs.gov/uac/SOI-Tax-Stats-Excise-Tax-Statistics [accessed September 2012].

Jaffe, A.B., and R.N. Stavins. 1994. Energy-efficiency investments and public policy. Energy J. 15(2):43-66.

JCT (Joint Committee on Taxation). 2008. Estimates of Federal Tax Expenditures for Fiscal Years 2008-2012. JCS-2-08. Prepared for the House Committee on Ways and Means and the Senate Committee on Finance, by the Staff of the Joint Committee on Taxation. Washington, DC: U.S. Government Printing Office [online]. Available: http://www.jct.gov/s-2-08.pdf [accessed June 6, 2013].

JCT (Joint Committee on Taxation). 2009. Carbon Audit of Provisions of the Internal Revenue Code of 1986 (sec.117 of the Act). Pp. 311-312 in General Explanation of Tax Legislation Enacted in the 110th Congress. JCS-1-09. Washington, DC: U.S. Government Printing Office [online]. Available: https://www.jct.gov/publications.html?func=startdown&id=1990 [accessed June 6, 2013].

JCT (Joint Committee on Taxation). 2012. Estimates of Federal Tax Expenditures for Fiscal Years 2011-2015. JCS-1-12. Prepared for the House Committee on Ways and Means and the Senate Committee on Finance, by the Staff of the Joint Committee on Taxation. Washington, DC: U.S. Government Printing [online]. Available: https://www.jct.gov/publications.html?func=startdown&id=4385[accessed June 6, 2013].

Jenkins, J., T. Nordhaus, and M. Shellenberger. 2011. Energy Emergence: Rebound & Backfire as Emergent Phenomena. Oakland, CA: Breakthrough Institute.

Jorgenson, D.W., and P.J. Wilcoxen. 1991. Reducing U.S. carbon dioxide emissions: The cost of different goals. Pp. 125-128 in Energy, Growth, and the Environment, J.R. Moroney, ed. Greenwich, CT: JAI Press.

Kasting, J.F., and D. Catling. 2003. Evolution of habitable planets. Annu. Rev. Astron. Astrophys. 41:429-463.

Keeney, R., and T.W. Hertel. 2009. The indirect land use impacts of U.S. biofuels policies: The importance of acreage, yield and bilateral trade responses. Am. J. Agr. Econ. 91(4):895-909.

Ko, J., and C. Dahl. 2001. Interfuel substitution in U.S. electricity generation. Appl. Econ. 33(14):1833-1843.

Krueger, A.B. 2009. Testimony of Alan B. Krueger, Assistant Secretary for Economic Policy and Chief Economist, U.S. Department of Treasury, before the U.S. Senate Committee on Finance, Subcommittee on Energy, Natural Resources and Infrastructure, September 10, 2009, Washington, DC.

Krupnick, A.J., I.W.H. Parry, M. Walls, T. Knowles, and K. Hayes. 2010. Toward a New National Energy Policy: Assessing the Options. Washington, DC: Resources for the Future.

Loulou, R., G. Goldstein, and K. Noble. 2004. Documentation for the MARKAL Family of Models. Energy Technology Systems Analysis Programme [online]. Available: http://www.iea-etsap.org/web/MrklDoc-I_StdMARKAL.pdf [accessed June 6, 2013].

Melillo, J.M., J.M. Reilly, D.W. Kicklighter, A. Gurgel, T.W. Cronin, S. Paltsev, B.S. Felser, X. Wang, A.P. Sokolov, and C.A. Schlosser. 2009. Indirect emissions from biofuels: How important? Science 326(5958):1397-1399.

Metcalf, G. 2007. Federal tax policy towards energy. Pp. 145-184 in NBER Tax Policy and the Economy, Vol. 21, J.M. Poterba, ed. Cambridge, MA: MIT Press.

Metcalf, G. 2009. Taxing Energy in the United States: Which Fuels Does the Tax Code Prefer? New York: Manhattan Institute.

Metcalf, G. 2010. Investment in Energy Infrastructure and the Tax Code. Pp. 1-33 in Tax Policy and the Economy, Vol. 24, J.R. Brown, ed. Chicago: University of Chicago Press.

Meyers, W., P. Westhoff, J. Fabiosa, and D. Hayes. 2010. The FAPRI global modelling system and outlook process. J. Int. Agr. Trade Dev 6(1):1-19.

Minnesota IMPLAN Group. 2008. State-Level U.S. Data for 2006. Stillwater, MN: Minnesota IMPLAN Group.

Mosnier, A., P. Havlík, H. Valin, J.S. Baker, B.C. Murray, S. Feng, M. Obersteiner, B.A. McCarl, S.K. Rose, and U.A. Schneider. 2012. The Net Global Effects of Alternative U.S. Biofuel Mandates: Fossil Fuel Displacement, Indirect Land Use Change, and the Role of Agricultural Productivity Growth. Report NI R 12-01. Nicholas Institute for Environmental Policy Solutions, Duke University [online]. Available: http://nicholasinstitute.duke.edu/climate/policydesign/net-global-effects-of-alternative-u.s.-biofuel-mandates [accessed June 6, 2013].

Nelson, R.R. 1959. The simple economics of basic scientific research. J. Polit. Econ. 67(3):297-306.

Newell, R., A. Jaffe, and R. Stavins. 1999. The induced innovation hypothesis and energy-saving technological change. Q. J. Econ. 114(3):941-975.

NRC (National Research Council). 1979. Carbon dioxide and climate: A scientific assessment, Washington, DC: National Academy Press.

NRC (National Research Council). 1983. Carbon Dioxide Assessment Committee. (1983). Changing Climate: Report of the Carbon Dioxide Assessment Committee, Washington, DC: National Academy Press.

NRC (National Research Council). 1992. Committee on Science, Engineering, and Public Policy, Policy Implications of Greenhouse Warming: Mitigation, Adaptation, and the Science Base, National Academy Press, Washington, D.C.

NRC (National Research Council). 2002. National Research Council, Committee on Abrupt Climate Change (2002), Abrupt Climate Change: Inevitable Surprises, National Academy Press, Washington, D.C.

NRC (National Research Council). 2010. Limiting the Magnitude of Future Climate Change, Washington, DC: National Academy Press.

NRC (National Research Council). 2011. Committee on America's Climate Choices, America's Climate Choices, Washington, DC: National Academy Press.

OMB (U.S. Office of Management and Budget). 2011. Estimate of total income tax expenditures for fiscal years 2010-2016. P. 241 in Fiscal Year 2012 Analytical Perspectives Budget of the U.S. Government [online]. Available: http://www.whitehouse.gov/sites/default/files/omb/budget/fy2012/assets/spec.pdf [accessed June 6, 2013].

Paltsev, S., J.M. Reilly, H.D. Jacoby, R.S. Eckaus, J. McFarland, M. Sarofim, M. Asadoorian, and M. Babiker. 2005. The MIT Emissions Prediction and Policy Analysis (EPPA) Model: Version 4. MIT Joint Program Report 125. Cambridge, MA: MIT [online]. Available: http://globalchange.mit.edu/files/document/MITJPSPGC_Rpt1 25.pdf [accessed June 6, 2013].

Paltsev, S., J.M. Reilly, H.D. Jacoby, and J.F. Morris. 2009. The costs of climate policy in the United States. Energ. Econ. 31(suppl. 2):S235-S243.

Plevin, R., M. O'Hare, A. Jones, M. Torn, and H. Gibbs. 2010. Greenhouse gas emissions from biofuels' indirect land use change are uncertain but may be much greater than previously estimated. Environ. Sci. Technol. 44(21):8015-8021.

Popp, D. 2002. Induced innovation and energy prices. Am. Econ. Rev. 92(1):160-180.

Popp, D. 2004. ENTICE: Endogenous technological change in the DICE model of global warming. J. Environ. Econ. Manage. 48(1):742-768.

Popp, D., and R.G. Newell. 2012. Where does energy R&D come from? A first look at crowding out from environmentally-friendly R&D. Energ. Econ. 34(4):980-991.

Popp, D., I. Hascic, and N. Medhi. 2011. Technology and the diffusion of renewable energy. Energ. Econ. 33(4):648-662.

Price, J. 2002. The production tax credit: Getting more credit than it's due? Fortnightly Magazine (May 15):38-41.

Rausch, S., G. Metcalf, J. Reilly, and S. Paltsev. 2010. Distributional implications of alternative U.S. greenhouse gas control measures. B.E. J. Econ. Anal. Poli. 10(2), DOI: 10.2202/1935-1682.2537.

Roberts, M., and W. Schlenker. 2010. The U.S. Biofuel Mandate and World Food Prices: An Econometric Analysis of the Demand and Supply of Calories. NBER Working Paper N 15921.

Romer, P. 1990. Endogenous technological change. J. Polit. Econ. 98(part 2):S71-S102.

Ross, M.T., A.A. Fawcett, and C.S. Clapp. 2009. U.S. climate mitigation pathways post-2012: Transition scenarios in ADAGE. Energ. Econ. 31(suppl. 2):S212-S222.

Rutherford, T.F. 1999. Applied general equilibrium modeling with MPSGE as a GAMS subsystem: An overview of the modeling framework and syntax. Comput. Econ. 14(1-2):1-46.

Santer, B.D., J.F. Painter, C.A. Mears, C. Doutriaux, P. Caldwell, J.M. Arblaster, P.J. Cameron-Smith, N.P. Gillett, P.J. Gleckler, J. Lanzante, J. Perlwitz, S. Solomon, P.A. Stott, K.E. Taylor, L. Terray, P.W. Thorne, M.F. Wehner, F.J. Wentz, T.M.L. Wigley, L.J. Wilcox, and C.Z. Zou. 2013. Identifying human influences on atmospheric temperature. Proc. Natl. Acad. Sci. USA 110(1):26-33.

Schäfer, A., J.B. Heywood, H.D. Jacoby, and I.A. Waitz. 2009. Transportation in a Climate-Constrained World. Cambridge: MIT Press.

Schnepf, R., and B.D. Yacobucci. 2013. Renewable Fuel Standard (RFS): Overview and Issues. CRS Report for Congress R40155. Congressional Research Service [online]. Available: http://www.fas.org/sgp/crs/misc/R40155.pdf [accessed June 6, 2013].

Searchinger, T., R. Heimlich, R. Houghton, F. Dong, A. Elobeid, J. Fabiosa, S. Tokgoz, D. Hayes, and T.H. Yu. 2008. Use of U.S. croplands for biofuels increases greenhouse gases through emissions from land-use change. Science 319(5867):1238-1240.

Small, K.A., and K. Van Dender. 2007. Fuel efficiency and motor vehicle travel: The declining rebound effect. Energy J. 28(1):25-52.

Stavins, R. 1995. Transactions costs and tradable permits. J. Environ. Econ. Manage. 29:133-148.

Taylor, L.D. 1975. The demand for electricity: A survey. Bell J. Econ. 6(1):74-110.

Taylor, L.D. 1977. The demand for electricity: A survey of price and income elasticities. In International Studies of the Demand for Energy, W.D. Norhaus, ed. Amsterdam: North Holland.

Texas Comptroller of Public Accounts. 2008. Government financial subsidies. Pp. 367-402 in The Energy Report, May 2008 [online]. Available: http://www.window.sta te.tx.us/specialrpt/energy/subsidies/ [accessed September 13, 2012].

Thompson, W., S. Meyer, P. Westhoff. 2008. Model of the U.S. Ethanol Market. FAPRI-MU Report 07-08. Food and Agricultural Policy Research Institute (FAPRI), University of Missouri-Columbia [online]. Available: https://mospace.umsystem.edu/xmlu i/bitstream/handle/10355/2663/modelofUSEthanolMarket.pdf?sequence=1 [accessed June 6, 2013].

Thompson, W., S. Meyer, and P. Westhoff. 2010. The new markets for renewable identification numbers. Appl. Econ. Perspect. Pol. 32(4):588-603.

Thompson, W., J. Whistance, and S. Meyer. 2011. Effects of U.S. biofuel policies on U.S. and world petroleum product markets with consequences for greenhouse gas emissions. Energ. Policy 39(9):5509-5518.

Tyner, W.E., F. Taheripour, Q. Zhuang, D. Birur, and U. Baldos. 2010. Land Use Change and Consequent CO_2 Emissions due to U.S. Corn Ethanol Production: A Comprehensive Analysis. U.S. Department of Agricultural Economics, Purdue University [online]. Available: https://ethanol.org/pdf/contentmgmt/Purdue_new_ILUC_report_April_2010.pdf [accessed June 6, 2013].

USDA (U.S. Department of Agriculture). 2011. World Agricultural Supply and Demand Estimates (WASDE) [online]. Available: http://www.usda.gov/oce/commodity/wasde/ [accessed June 6, 2013].

Wade, S.H. 2003. Price Responsiveness in the AEO2003 NEMS Residential and Commercial Buildings Sector Models. Energy Information Administration [online]. Available: ftp://ftp.eia.doe.gov/forecasting/analysispaper/buildings.pdf [accessed June 6, 2013].

Westhoff, P., J. Binfield, and S. Gerlt. 2012. U.S. Baseline Briefing Book: Projections for Agricultural and Biofuel Markets. FAPRI-MU Report #01-12. Food and Agricultural Policy Research Institute, University of Missouri [online]. Available: http://www.fapri.missouri.edu/outreach/publications/2012/FAPRI_MU_Report_01_12.pdf [accessed June 6, 2013].

Wise, M., K. Calvin, A. Thomson, L. Clarke, B. Bond-Lamberty, R. Sands, S.J. Smith, A. Janetos, and J. Edmonds. 2009. Implications of limiting CO_2 concentrations for land use and energy. Science 324(5931):1183-1186.

Wiser, R. 2007. Wind Power and the Production Tax Credit: An Overview of Research Results. Testimony Prepared for a Hearing on Clean Energy by the U.S. Senate Finance Committee, March 29, 2007, Washington, DC [online]. Available: http://eetd.lbl.gov/ea/emp/reports/wiser-senate-test-4-07.pdf [accessed June 6, 2013].

Appendix A

Modeling Approaches to the Effects of Tax Policy on GHG Emissions

INTRODUCTION

This technical appendix is intended to assist those interested in examining the modeling approaches selected by the committee from among the alternatives in greater depth and detail and in understanding better their differences, strengths, and limitations. It compares the structures of available models and provides sources of documentation of the National Energy Modeling System for the National Academy of Sciences, Food and Agricultural Policy Research Institute at the University of Missouri (FAPRI-MU), Intertemporal General Equilibrium Model (IGEM), and Center for Business and Economic Research (CBER) models, whose results are described in the report.

Modeling approaches relevant to the questions before the committee range from those that compare the cost of alternative investment decisions, taking into account how taxes affect the investment return, to models of the entire economy that project how a tax or tax expenditure affects domestic and international markets and the overall level of economic activity. In between are models that focus on some limited set of markets that are directly affected by a tax provision or that interact strongly with markets that are directly affected. Narrow tax provisions that affect very specific investment decisions such as those directed toward wind turbine or solar photovoltaic installations require a detailed investment-level evaluation. The effect of the provision on cost of the project may depend on eligibility criteria, profitability of the investor if the value is as a credit against taxes, expected prices of inputs, and expected inflation. Whether the tax provision is actually used and has an impact on emissions will depend on how it alters the cost of the project and investment decisions.

Most sector or economy-wide models do not represent the structural details of investment decisions at the detailed level, but instead simplify the incentive effect as either an effective tax rate on investment, as a change in the levelized cost of the technology, or as a supply shift. Thus, a detailed model of investment decisions may be a needed first step in using a broader market model that must model an intricate tax provision as, for example, a simpler effective tax rate on a particular technology, sector, or factor input. The markets directly affected by the narrow provisions identified by the panel were energy markets, and for biofuel-related tax expenditures, agricultural markets. Energy and agricultural markets are themselves fairly complex, with a variety of existing regulatory policies that affect them and that potentially interact with tax incentives. The broad provisions the committee identified, such as accelerated depreciation or those related to housing and health care, require an economy-wide approach, or at least a scope beyond just energy and agricultural markets. In short, no single model was likely to have detail on agricultural markets and energy markets, while also capturing economy-wide effects of broad policies.

The committee's review focused on three types of models. First was a set of economy-wide models, often with some detail on energy or agricultural sectors. A second set of models has been developed with a strong focus on agricultural markets and the effect of biofuel policies on them, with varying degrees of detail on how biofuels would also affect energy markets. A third set of models has focused on energy markets in considerable detail. In principle, the set of economy-wide models are potentially capable of analyzing many or most of the tax provisions, but they are limited in that they generally lack the granularity needed for some of the detailed provisions. For example, an economy-wide model that represents the electricity sector as single production function cannot easily represent the effect of a provision directed just at wind, solar, or nuclear. And similarly, a model that simplifies the agricultural sector as producing an aggregate crop or livestock product is less able to trace how a biofuels policy may affect corn production and land-use change. For each of the tax provisions to be examined by one or more models, the first step is to determine the effect of the provisions on parameters available in the model, based on a detailed model of investment decisions, if needed.

ECONOMY-WIDE MODELS

The committee identified six economy-wide models with the capability to examine at least some of the tax provisions. These included (1) the MIT Emissions Predictions and Policy Analysis (EPPA) model (Paltsev et al., 2005, 2009); and/or (2) the MIT U.S. Regional Energy Policy (USREP) model (Rausch et al., 2010) very similar to EPPA but with greater detail on the United States; (3) the Applied Dynamic Analysis of the Global Economy (ADAGE) model (Ross et al., 2009) developed at RTI International and widely used by the Environmental Protection Agency (EPA) for analysis of greenhouse gas (GHG) policies (e.g., EPA, 2009); (4) the Multi-Region National (MRN) model developed at Charles River Associates (Bernstein et al., 2007); (5) the Global Trade Analysis Project (GTAP) model developed at Purdue University (Hertel et al.,

2010); and (6) the Intertemporal General Equilibrium Model (IGEM) of the United States (Goettle et al., 2007) developed at Dale Jorgenson Associates.

These models have some similarities and differences. All are multisectoral general equilibrium models that represent the economy following modern neo-classical economic theory, meaning that consumers and producers are both assumed to utility- or profit-maximize, given constraints. All are constructed on some version of input-output tables for the United States (or other regions of the world if included) and an expanded social accounting matrix that includes estimates of factor returns from each production sector and the disposition of goods to final-demand sectors (households, government, investment, and exports). Thus, a general strength of these models is that any ripple effects on other sectors, final demands, and exports and imports of a tax provision that affects one sector will be included in estimates of their greenhouse gas emissions effects. If changes in the energy sector affect the level of steel production or what type of energy it uses, these models will, in principle, include this effect. While some of these models cover the whole world, and some just the United States, they incorporate some estimate of effects of trade via export demand and import supply functions. They can also estimate how a tax may affect the overall level of economic activity via incentives for labor and investment.

Beyond these general theoretical similarities there are various similarities among some of them in terms of specific databases, solution approaches, and representation of dynamics, but this is also where some differences emerge. EPPA, ADAGE, and GTAP were developed using the same global economic database that is maintained and updated in the Global Trade Analysis Project at Purdue University. The USREP and MRN models were developed using the IMPLAN state-level economic database for the United States (Minnesota IMPLAN Group, 2008), and as with USREP, there is a companion global model to the MRN built on the GTAP database. The EPPA, USREP, MRN, and ADAGE models all utilize the General Algebraic Modeling System's (GAMS) Mathematical Programming System for General Equilibrium Analysis (MPSGE) model development and solver software (Rutherford, 1999). The GTAP model uses its own solution algorithm that allows more flexibility in the functional forms than does the GAMS/MPSGE approach. All of these are simulation models where the model developers have surveyed the literature for the value of critical parameters in the model. The key parameters are elasticities of substitution among inputs, and these generally determine how a change in a tax provision will ultimately affect markets. Here the IGEM model differs from the other models, as it includes a time series of data for the U.S. economy, and the parameters of response are econometrically estimated from these data. It also utilizes its own solution algorithm. Different appraisals of the literature, and differences from the econometric estimates of the IGEM model, mean these models will respond differently to tax provisions.

The models also differ in how they deal with investment and savings. The GTAP model is a static model. It is developed for the base-year data, and simulates how the economy would have been different in that year under different conditions, in our case, for example, the difference with or without a tax provision. In GTAP, capital is fixed and investment is based on a fixed savings rate. The EPPA and USREP models, in their standard formulations, are recursive dynamic models, meaning that investment is determined by a fixed savings rate. Unlike a static model, investment in one period becomes new capital in the next. ADAGE, MRN, and IGEM are forward-looking models, meaning that the savings rate is determined by the model. Higher returns to capital in the future lead to more savings by households today. This requires agents to look forward and anticipate the future returns to capital, making decisions today based on those expectations. Agents are said to have perfect foresight in these models because they are solved so that their expectations are realized exactly.

The models also deal with existing capital differently. The EPPA and USREP models have explicit capital vintaging by model sector, so once a factory or power plant of a particular vintage is put in service in a particular sector it is difficult to adjust it until it is fully depreciated. This is sometimes referred to as a putty-clay model: Capital is malleable when the investment is put in place, but once it is built, its features are fixed. ADAGE, MRN, and IGEM have simpler structures, generally some version of putty-putty, where capital is fully malleable and can be redeployed to other sectors. In some cases the current capital stock characteristics are fixed until it fully depreciates, but any future investment remains fully malleable over its lifetime.

These features can be important in the assessment of tax provisions: In provisions with sunset laws, one can observe forward-looking behavior, as there is sometimes a rush of investment to take advantage of a provision before it expires. Hence, both the forward-looking aspects of models and the capital vintaging can be important. Investment installed in anticipation of a sunset provision (e.g., a wind turbine) will remain as a viable contributor to the capital stock for 20 years or more. However, one can observe that often these provisions have been extended just as they are to expire, and so agents observing these are likely to have imperfect expectations—based on past experience they may give some chance that the provision will continue indefinitely. Agents also do not have a crystal ball into the future of energy prices, broader carbon policy, or other environmental policy. Explicitly modeling imperfect expectations or a fully stochastic solution to these models is not numerically feasible, and so forward-looking models certainly overestimate the capability of agents to look forward, even as recursive dynamic models may underestimate this ability.

There are also differences in the sectoral disaggregation and attention in these models. The EPPA, USREP, MRN, and ADAGE models have retained a relatively high level of aggregation for most sectors of the economy—a total of 6-10—but focused detail on the energy sector, including explicit representation of electricity-generation alternatives, transportation fuels, and vehicle options. The GTAP model has focused heavily on detail in the agriculture sector.

IGEM's model strategy is greater general disaggregation of production sectors, but within the constraints of the Standard Industrial Classification system. In this classification, for example, electricity generation is a sector, but there is no further distinction of solar, wind, nuclear, or hydro from generation using coal, gas, or oil. Fuels, capital, labor, and intermediate goods are inputs into the electricity sector, and some amount of electricity is produced. Implicitly, substitution of capital for fuels could be interpreted as an increase in one of the non-fossil-fuel technologies. Rates of technical change are econometrically estimated based on the historical data, and these are represented as time trends on input requirements rather than explicit technologies.

Along with the features discussed above, the IGEM model has the most extensive representation of the general tax system, a feature that was important for consideration of the broad provisions. IGEM is not as rich in energy or agriculture detail, and hence was not able to simulate many of the narrow provisions of interest to the committee, but its strength was potentially in the broad provisions. Its treatment of the effects of accelerated depreciation was a particular strength. Although IGEM is not as strong in analyzing the health and housing provisions, the committee found no model capable of treating these better for our purposes. With regard to treatment of housing provisions, the IGEM and other models above generally treat housing as household investment, and the rental value is a substitute for other household goods. For analysis of energy implications of change in investment of housing, it is more realistic to treat the investment level in housing as a complement to energy use rather than as a substitute. With less investment in housing, we would expect to see smaller houses, larger households and thus fewer houses, and so generally lower energy requirements and lower GHGs. As discussed in the main text, as far as we could ascertain there are no models capable of assessing the GHG impacts of specific tax policies that treat household capital and energy as complementary inputs, and so our ability to assess accurately the effects of tax provisions on housing was limited. IGEM simulated the impact of the tax treatment of health insurance as an untaxed portion of wages (supplied in the form of health insurance) rather than a subsidy to health care provision. Thus, it did not have any impact on the pricing of health care or the purchase decisions among goods by households. One might expect that if health insurance was more expensive because it was taxed like other income, that households would consume less health care and more of other goods, and so to the extent health services had different GHG implications than other goods we would see an effect. IGEM treatment of the housing and health care tax provisions is better suited to estimating how these exclusions distort investment decisions, and thus the level of overall economic activity, rather than how these provisions distort choices *among* goods and services.

It is difficult to assess how IGEM features might lead to different results than other models of this type, especially without simulating them. Even with comparable simulations, any particular outcome comes about as a result of hundreds of different parameters and a variety of complex differences in model structure. In general, we expect, and some comparisons have shown, that the forward-looking behavior results in much greater flexibility, as does the malleability of capital.

ENERGY-FOCUSED MODELS

There were three principal energy-related models considered by the committee. These were the MARKAL model (MARKAL) (Loulou, et al., 2004), which is more of a modeling framework that can be developed for specific applications depending on interests and data availability; the CBER model (Allaire and Brown, 2011), developed at the University of Nevada, Las Vegas; and the NEMS model, developed and used by the U.S. Energy Information Administration (EIA) in its annual energy outlook and available for other uses through On Location, Inc.

MARKAL energy models have been developed for individual energy sectors such as transportation or power generation, or for the entire energy system for many different countries. MARKAL models are engineering cost-based models, where technological options are presented as fixed coefficient production technologies. They generally take prescribed energy or service demands as fixed and explicitly minimize the cost of meeting those demands. They are technology rich. Preferences among technologies are fully determined by out-of-pocket (pecuniary) costs. This assumption may be reasonable for large-scale investments in power plants, where a relatively few well-described options exist and local conditions that are poorly known are unlikely to affect the cost ranking of different options. However, the ability of the approach to realistically simulate outcomes rests on its ability to fully describe regional considerations that might affect choices, and all the options available. Transportation demand for fuel is an example of some of the limits of this approach. Broad vehicle options are represented with different costs and energy-use requirements, and change in the choice among these options is the only possible response to changing fuel price. But a MARKAL model has difficulty representing the near-continuous nature of actual or potential options in terms of transmission, engine, size, weight, and energy-consuming options that are available, or could be offered if there was demand. Miles driven are prescribed and unaffected by fuel price. Decisions about maintenance or how one drives that might affect fuel use are also assumed to be unaffected by price. Preferences for amenities of vehicles and willingness to trade these against vehicles with greater fuel efficiency are not represented. Thus, this class of models tends to "over optimize," represent stark technology options, and not consider any behavior or preference other than strictly minimizing vehicle and fuel cost without consideration of other features of the technology that may deliver welfare value to consumers. In this type of model, the purchase of expensive, relatively low-mileage luxury vehicles would appear irrational, whereas an econometric approach based on observed behavior would show resistance to giving up luxury for fuel economy.

The CBER model is a polar opposite of MARKAL-type models. In the CBER model, demand and supply for energy are represented as continuous functions. The key parameters in the CBER model are price elasticities of supply and demand. Elasticities for CBER are deduced from reviews of econometric literature. The strength of such an approach is that implicitly regional variation in the cost of technology, consumer's willingness to make trade-offs, and broad technological differences are captured to the extent these differences were captured in the historical data. The astructural nature of the model means that it may not capture dynamics of capital turnover well, and it is somewhat captive of the history in terms of the available technology and the structure of policy. For example, the introduction of the Renewable Fuels Standard (RFS) for biofuels substantially changes the fuel market, a relatively recent development that is not reflected in historical supply-demand data on which CBER-type models rely for estimates of elasticities.

An advantage of the CBER model is that it was developed with an objective of investigating the effects of tax policy on GHG emissions in the energy sector, and is the only significant effort in this regard to have been done prior to the committee's work. Given that focus, the developers have invested effort in representing most of the energy-specific tax code provisions. That said, because the model lacks the specific structure around which the tax code affects decisions—tax code changes are translated into simple shifts in demand for supply of a fuel or electricity—the outcomes are highly dependent on the analysis and availability of data, or lack thereof, that goes into determining how supply or demand are shifted by a tax code provision. In contrast, all of the other models we evaluated were more structural—a change in the tax code could be represented as a change in the cost of capital, for example.

The NEMS model is an eclectic combination of the previous two types of models with considerable detail on the structure of markets and policies affecting them and, in general, far more sectoral demand and energy-supply detail. The power sector is similar to a MARKAL-type model, but other sectors of the economy are represented by demand and supply elasticities. It has benefitted from a large and ongoing investment in it to keep up with changes in policy and technology options because results of each EIA annual energy outlook are carefully scrutinized. While it can be run in conjunction with a macroeconomic model with iteration to achieve energy-sector macroeconomic feedbacks, that was not an option that was feasible in applications considered here. Thus, aggregate economic and industrial activity were held constant. And those feedbacks between the energy sector and the rest of the economy are not as fully integrated as in the macroeconomic models discussed above.

In the end, the committee was able to commission additional runs from the CBER model and runs of the NEMS model, conducted by On Location, Inc. The strength of the CBER approach is in the breadth of tax provisions it was able to consider. Its weakness was lack of structural detail—it does not explicitly represent existing policies such as the RFS2, capital as different from other inputs, or the relationship of the energy sector to the macroeconomy. The advantage of the NEMS is in both technological and market structural detail on the energy sector. Its weakness is lack of macroeconomic feedback, and more limited detail in the agricultural sector that is important for biofuels tax provisions. However, it does include important structural and policy details of fuels markets and the RFS2 provisions.

AGRICULTURE-SECTOR MODELS

The committee's interest in agriculture-sector models stemmed largely from tax provisions related to biofuels. A large variety of approaches have been used to examine the economics of biofuels and biofuel policy. These include macroeconomic models such as EPPA and GTAP (e.g., Gurgel et al., 2007, 2011; Tyner et al., 2010; Decreux and Valin, 2007) broadly reviewed above, agricultural optimization models including FASOM (Adams et al., 1996, as in Beach and McCarl, 2010; Beach et al., 2010), simulation models such as MiniCAM (Wise et al., 2009), and econometric-based simulation models such as FAPRI (Babcock and Carriquiry, 2010). There are several challenging aspects of modeling biofuel policy, including (1) the complex interactions with agriculture and agricultural policy, including competing demands for crops and by-products supplies of animal feeds; (2) the complex policy requirements of the Renewable Fuel Standard (RFS2, as described below) and investment and production tax credits that differentially treat different biofuel production pathways and feedstocks that are the focus of this report; (3) international linkages in agriculture and energy markets; (4) land-use change and competition for land; and (5) the carbon implications of land-use change. The macroeconomic models are generally more coarsely resolved and have less detail on the agriculture sector and less explicit representation of policy; however, the GTAP model represents agriculture in considerable detail. A limit, as discussed above, is that it is purely static. FASOM has some similarities to a MARKAL-based energy model that chooses least cost production activities. Given the importance of regional differences, it is explicitly spatial, and it represents demands as continuous functions, where key parameters are demand elasticities. FAPRI at its base is similar to the CBER model, but incorporates considerably more structural detail on policy as it affects markets, and approximates dynamic adjustment of markets over time. While it is not maintained by a federal agency as is NEMS, the FAPRI model produces an annual agricultural baseline projection that incorporates changing policy in the agricultural sector. While the FASOM and FAPRI models were initially purely agricultural-sector models, given the interest in biofuels, they have added much more detail on gasoline and petroleum markets and thus are in better position to analyze these interactions.

The FAPRI-MU model as applied in this study is a system of demand and supply functions for 16 crops, 15 crop-based products, and 17 different types of livestock and livestock-based products (Meyers et al., 2010; Devadoss et al., 1993). Some equations are econometrically estimated, others are not. In particular, the rapid changes in biofuel markets make direct estimation of biofuel-related equations based on observed behavior problematic. A good example is E85 demand, which can be very important in projections, but has not accounted for more than a very small amount in the past. In contrast, FAPRI-MU updates estimates for equations in some other components where there are more historical data.

The model's focus is on the United States, but the rest of world is either collapsed into a single rest-of-world supply-and-demand response as in the case of animal products, composed of a similar rest-of-world aggregate response but with key countries identified, as in the case of ethanol, or represented with aggregate rest-of-world supply-and-demand aggregates, as in the case of main crops. This longstanding agricultural model has been recently augmented to include detailed modules on oil markets (Thompson et al., 2011) and U.S. biofuels markets (Thompson et al., 2008). The strength of the model is in its detailed representation of agricultural markets, including global markets, modeling of the complex Renewable Identification Number fuel credits with multiple fuel production pathways representing both conventional and second-generation processes, and links to global petroleum and refined fuel markets (Thompson et al., 2010). The modeling approach does not explicitly consider land use or the carbon implications of land-use change. These are highly uncertain responses with wide-ranging results in the literature (Plevin et al., 2010; Searchinger et al., 2008; Melillo et al., 2009; Keeney and Hertel, 2008; Hertel, 2011; Tyner et al., 2010; Mosnier et al., 2012). Instead, greenhouse gas implications are assessed by applying a fixed GHG coefficient per unit of fuel for different biofuel production pathways. Default estimates are those of the U.S. EPA (2010) that include CO_2, N_2O, and CH_4 implications of land-use change.

DOCUMENTATION OF THE MODELS USED BY THE COMMITTEE

Intertemporal General Equilibrium Model (IGEM). Documentation is available at http://www.igem.insightworks.com/docs/190.html.

National Energy Modeling System (NEMS). Documentation is available on the website of the U.S. Energy Information Administration at http://www.eia.gov/analysis/model-documentation.cfm (various modules) and http://www.eia.gov/forecasts/nemsdoc/integrating/pdf/m057(2012).pdf (integrating mo-dule, 08/2012).

Center for Business and Economic Research Model (CBER). The equations used in the CBER analysis can be found in Appendix A of the report by Allaire and Brown: Allaire, M., and Brown, S. (August 2012). *U.S. Energy Subsidies: Effects on Energy Markets and Carbon Dioxide Emissions.* Retrieved 2012, from http://www.pewtrusts.org/uploadedFiles/wwwpewtrustsorg/Reports/Fiscal_and_Budget_Policy/EnergySubsidiesFINAL.pdf.
Allaire and Brown and Brown and Kennelly used the MATLAB program to solve the system of equations within the CBER model. Any mathematical solver package, including open-source solvers, can be used to solve the equations and thus replicate results of the CBER model.

Food and Agriculture Policy Research Institute Model (FAPRI). Documentation is available at http://www.fapri.missouri.edu in the following reports:

1. FAPRI-MU Report #12-11, Model Documentation for Biomass, Cellulosic Biofuels, Renewable and Conventional Electricity, Natural Gas and Coal Markets
2. FAPRI-MU Report #09-11, FAPRI-MU Stochastic U.S. Crop Model Documentation
3. FAPRI-MU Report #05-11, New Challenges in Agricultural Modeling: Relating Energy and Farm Commodity Prices
4. FAPRI-MU Report #09-10, FAPRI-MU U.S. Biofuels, Corn Processing, Distillers Grains, Fats, Switchgrass, and Corn Stover Model Documentation
5. FAPRI-MU Report #07-08, Model of the U.S. Ethanol Market FPARI-UMC Report #12-04, Documentation of the FAPRI Modeling System Consultant Reports Detailing Results of Modeling Efforts
6. The original modeling using the models described above was undertaken by four independent consultants. Each of those consultants produced reports to the committee detailing the results of their modeling efforts. Readers can download those reports at the National Academies Press website, http://www.nap.edu/catalog.php?record_id=18299.

Appendix B

Biographical Information of Committee and Staff

WILLIAM D. NORDHAUS is Sterling Professor of Economics at Yale University, New Haven, Connecticut, USA. He received a B. A. from Yale University in 1963 and a Ph.D. in Economics in 1967 from the Massachusetts Institute of Technology.

Nordhaus is a member of the National Academy of Sciences, a Fellow of the American Academy of Arts and Sciences, and a Fellow of the American Philosophical Society. He is on the research staff of the National Bureau of Economic Research and the Cowles Foundation for Research. From 1977 to 1979, he was a Member of the President's Council of Economic Advisers. From 1986 to 1988, he served as the Provost of Yale University. Professor Nordhaus is current or past associate editor of several scientific journals. He was elected President of the American Economic Association to serve in 2014-2015.

He has served on several committees of the National Academy of Sciences on topics including climate change, environmental accounting, and the macroeconomic effects of aging. He is the author of many books and scholarly publications. His books include *Is Growth Obsolete?* (with James Tobin), *The Efficient Use of Energy Resources*, *Reforming Federal Regulation*, *Managing the Global Commons*, and (joint with Paul Samuelson) the classic textbook, *Economics*, now in its nineteenth edition. He is author of a book on the economics of global warming, *The Climate Casino,* to be published by Yale Press in 2013.

MAUREEN L. CROPPER is professor of economics at the University of Maryland, former lead economist at The World Bank, and research associate at the National Bureau of Economic Research. Dr. Cropper's research has focused on valuing environmental amenities (especially environmental health effects), on the discounting of future health benefits, and on the tradeoffs implicit in environmental regulations. Her recent research analyzes the externalities associated with motorization and the interaction between residential location, land use and travel demand. Dr. Cropper is a member of the National Academy of Sciences. She also is past president of the Association of Environmental and Resource Economists and a former chair of the Advisory Council for Clean Air Act Compliance Analysis, a subcommittee of Environmental Protection Agency's Science Advisory Board. Dr. Cropper has served on the advisory boards of Resources for the Future, the Harvard Center for Risk Analysis, the Donald Bren School of the Environment and the AEI-Brookings Center on Regulation. She received her Ph.D. in economics from Cornell University.

FRANCISCO DE LA CHESNAYE is a Program Manager and Senior Economist at the Electric Power Research Institute (EPRI). He manages the institute's program on Global Climate Policy Costs and Benefits and also manages the application of a new U.S. energy-economic model used to assess the impact of climate and energy policies on the electric power sector, the energy system, and the economy at both regional and national scales. In addition to his work at EPRI, Mr. de la Chesnaye serves on various external expert panels. He is a Lead Author on the current U.S. Climate Assessment's Mitigation chapter. In 2010, Mr. de la Chesnaye served on a previous Academies Panel which completed a report to Congress titled "Limiting the Magnitude of Future Climate Change." Prior to joining EPRI in 2008, Mr. de la Chesnaye was the Chief Climate Economist at the U.S. Environmental Protection Agency where he was responsible for developing and applying the agency's economic models for domestic and international climate change policy analysis for the Administration and Congress. Mr. de la Chesnaye was a Lead Author for the Intergovernmental Panel on Climate Change's Fourth Assessment Report and was a co-editor of Human-Induced Climate Change: An Interdisciplinary Assessment published by Cambridge Univ. Press (2007). Mr. de la Chesnaye is a Ph.D. candidate at the Univ. of Maryland's School of Public Policy. He holds graduate degrees from Johns Hopkins and American University and a B.S. from Norwich Univ., The Military College of Vermont.

NOAH DIFFENBAUGH is an Assistant Professor in the School of Earth Sciences and Center Fellow in the Woods Institute for the Environment at Stanford University. His research interests are centered on the dynamics and impacts of climate variability and change, including the role of humans as a coupled component of the climate system. Much of his work has focused on the role of fine-scale processes in shaping climate change impacts, including studies of extreme weather, water resources, agriculture, human health, and poverty vulnerability. Dr. Diffenbaugh is currently a Lead Author for Working Group II of the Intergovernmental Panel on Climate Change (IPCC). He also serves on the Executive Committee of the Atmospheric Sciences Section of the American Geophysical Union (AGU), as an Editor of Geophysical Research Letters, and as a Member Representative to the University Corporation for Atmospheric Research (UCAR). Dr. Diffenbaugh is a recipient of the James R. Holton Award from the

American Geophysical Union, a CAREER award from the National Science Foundation, and a Terman Fellowship from Stanford University. He has also been recognized a Kavli Fellow by the U.S. National Academy of Sciences, and as a Google Science Communication Fellow. Before coming to Stanford, he was a member of the faculty of Purdue University, where he was a University Faculty Scholar and served as Interim Director of the Purdue Climate Change Research Center (PCCRC).

DAVID G. HAWKINS is Director of the Natural Resources Defense Council (NRDC) Climate Programs, which focus on advancing policies and programs to reduce pollution responsible for global warming and harmful climate change. He has also been director of the NRDC Air and Energy Program and was co-director of the NRDC Clean Air Program. He initiated the NRDC Clean Air Project in 1971, which has influenced the federal Clean Air Act since the law's passage. He served as Assistant Administrator for Air, Noise, and Radiation at EPA from 1977 to 1981. He served as a member of the National Research Council Board on Energy and Environmental Systems and committees to Review the Structure and Performance of the Health Effects Institute and to study the Energy Futures and Air Pollution in Urban China and the United States. Mr. Hawkins has extensive public policy and regulatory experience related to air quality, climate change, and related energy supply and demand issues.

ROBERTA MANN is the Frank Nash Professor of Law at the University of Oregon School of Law. A recognized tax law expert, she has written extensively on how tax policy affects the environment. Professor Mann earned her B.S., M.B.A. and J.D., cum laude, from Arizona State University, where she also served as assistant editor of the Arizona State University Law Journal. In 1995, she received her LL.M. in taxation, with distinction, from Georgetown University Law Center. Prior to entering academia, Professor Mann practiced in the Office of Chief Counsel of the Internal Revenue Service, concentrating primarily on the areas of partnerships, corporate, and natural resources. She also served on the Staff of the Joint Committee on Taxation in U.S. Congress.

BRIAN MURRAY is director of economic analysis at Nicholas Institute for Environmental Policy Solutions. Dr. Murray is widely recognized for his work on the economics of climate change policy, including the design of cap-and-trade policy elements to address cost containment and inclusion of offsets from traditionally uncapped sectors such as forestry and agriculture. His work has also focused on the broader economic and environmental implications of policies to expand biofuel use. He routinely advises members of the United States Congress and their staff on climate change legislative proposals. Dr. Murray has been invited as a co-author of several national and international assessments of forest resources, especially related to climate change. Of particular note, he was a convening lead author of the Intergovernmental Panel on Climate Change's Special Report on Land Use, Land Use Change, and Forestry. He has convened several forums of economic modeling experts to examine and communicate the results of their climate, energy and land use policy efforts to the public and private sectors. His research has examined the economic effects of traditional command-based regulatory strategies for pollution control and more market-oriented approaches such as cap-and-trade programs and emission taxes. His work has been published in books, edited volumes, and professional journals. Prior to coming to the Nicholas Institute in 2006, Dr. Murray was Director of the Center for Regulatory Economics and Policy Research at RTI International, a university-affiliated not-for-profit research institution.

JOHN M. REILLY is Co-Director of MIT's Joint Program on the Science and Policy of Global Change and a Senior Lecturer in the Sloan School. He is an energy, environmental, and agricultural economist who focuses on understanding the role of human activities as a contributor to global environmental change and the effects of environmental change on society and the economy. A key element of his work is the integration of economic models of the global economy as it represents human activity with models of biophysical systems including the ocean, atmosphere, and terrestrial vegetation. By understanding the complex interactions of human society with our planet, the goal is to aid in the design of policies that can effectively limit the contribution of human activity to environmental change, to facilitate adaptation to unavoidable change, and to understand the consequences of the deployment of large scale energy systems that will be needed to meet growing energy needs.

DREW SHINDELL is a senior scientist at the National Aeronautics and Space Administration's Goddard Institute for Space Studies. Dr. Shindell researches climate change, with a focus on atmospheric chemistry. An expert on modeling the impact of emissions changes, Dr. Shindell's work has investigated how the atmospheric chemical system has important effects on humans through pollutants such as smog or particulates, through acid rain, and through stratospheric ozone change, and how climate can be altered by greenhouse gases, solar variability, volcanic eruptions, aerosols, and ozone, and what impacts changes in climate and air quality may have on society. Dr. Shindell serves as a coordinating lead author for the Intergovernmental Panel on Climate Change's Fifth Assessment Report on global climate change. He earned his Ph.D. at Stony Brook University.

ERIC TODER is an Institute Fellow at the Urban Institute and co-director of the Urban-Brookings Tax Policy Center. Dr. Toder's recent work includes papers on using a carbon tax to pay for corporate rate cuts, cutting tax preferences to pay for lower tax rates, tax expenditures and the size of government, tax policy and international competitiveness, value added taxes, the home mortgage interest deduction, trends in tax expenditures, the distributional effects of tax expenditures, corporate tax reform, charitable tax incentives, taxation of saving, the tax gap, effects on retirement income of changes in pension coverage and stock prices, employing older workers, and energy tax incentives. Dr. Toder previously held a number of positions in tax policy offices in the U.S. government and overseas, including service as Deputy Assistant Secretary for Tax Analysis at the U.S. Treasury Department, Director of Research at the Internal Revenue Service, Deputy Assistant Director for Tax Analysis at the Congressional Budget Office, and consultant to the New Zealand Treasury. He received his Ph.D. in economics from the University of Rochester in 1971.

ROBERTON C. WILLIAMS, III studies both environmental policy and tax policy, with a particular focus on interactions between the two. He is an associate professor at the University of Maryland, College Park, senior fellow and director of academic affairs at Resources for the Future, and a research associate of the National Bureau of Economic Research. He also serves as a co-editor of the Journal of Public Economics, editorial council member (and former co-editor) of the Journal of Environmental Economics and Management and member of the editorial board of the B.E. Journal of Analysis & Policy. He was previously an associate professor at the University of Texas at Austin, a visiting research scholar at the Stanford Institute for Economic Policy Research, and an Andrew W. Mellon fellow at the Brookings Institution.

CATHERINE WOLFRAM is the Flood Foundation Professor of Business Administration at the Haas School of Business at the University of California, Berkeley. She also serves as co-director of the Energy Institute at Haas and is a member of the Haas School's Economic Analysis & Policy Group. Professor Wolfram studies the economics of energy markets and has examined the impact of environmental regulation on energy markets and the effects of electricity industry privatization and restructuring around the world. She earned her Ph.D. in economics from the Massachusetts Institute of Technology.

STAFF

STEPHEN A. MERRILL, project director, has been Executive Director of the National Academies' Board on Science, Technology, and Economic Policy (STEP) since its formation in 1992. With the sponsorship of numerous federal government agencies, foundations, multinational corporations, and international institutions, the STEP program has become an important discussion forum and authoritative voice on innovation, competitiveness, intellectual property, human resources, statistical, and research and development policies. At the same time Dr. Merrill has directed many STEP projects and publications, including *A Patent System for the 21st Century* (2004), *Innovation Inducement Prizes* (2007), and *Innovation in Global Industries* (2008). For his work on patent reform he was named one of the 50 most influential people worldwide in the intellectual property field by *Managing Intellectual Property* magazine and earned the Academies' 2005 Distinguished Service Award. He has been a member of the World Economic Forum Global Council on the Intellectual Property System. Previously, Dr. Merrill was a Fellow in International Business at the Center for Strategic and International Studies (CSIS), where he specialized in technology trade issues. He served on various congressional staffs including the U.S. Senate Commerce, Science, and Transportation Committee, where he organized the first congressional hearings on international competition in the semiconductor and biotechnology industries. Dr. Merrill holds degrees in political science from Columbia (B.A.), Oxford (M.Phil.), and Yale (M.A. and Ph.D.) Universities. He attended the Kennedy School of Government's Senior Executives Program and was an adjunct professor of international affairs at Georgetown University from 1989 to 1996.

PAUL BEATON first joined STEP in 2010 as a Mirzayan Science & Technology Policy Fellow and returned in 2011 as a Program Officer. He brings significant experience and expertise in entrepreneurship and innovation, and in science and technology policy, particularly in the areas of healthcare technology and energy and natural resources. At STEP he works on a spectrum of issues from tax and energy to immigration and intellectual property. Paul also serves on the advisory board of the African Institute for Health Policy and as an independent advisor to a healthcare IT company.

Prior to joining STEP, Paul earned a combined J.D. and Master of Environmental Management from the Yale Law School and the Yale School of Forestry & Environmental Studies, where he received the Benjamin F. Stapleton, Jr. Scholarship, a Dean's Scholarship, and a Sonnenschein Scholarship. At Yale he focused primarily on energy and climate change law and policy, authoring research on the climate effects of U.S. agriculture policy and an analysis of policy and structure at the United Nations Intergovernmental Panel on Climate Change. Paul also worked with a negotiating team and heads of state at United Nations' climate change treaty negotiations, and has counseled on Latin American nations on environmental legislation. He also served on the executive board of *The Yale Journal on Regulation* and as an executive officer of the Yale Environmental Law Association.

Paul earned his B.Sc. (*summa cum laude*) from UNC-Asheville where he received a congressionally awarded Barry M. Goldwater Scholarship, the William and Ida Friday Award, and was elected to Omicron Delta Kappa. Paul also received the University Fellows Grant, which along with a NSF grant and his Goldwater Scholarship, supported his research on the impacts of industrial substances that have replaced ozone-depleting chemicals since the enactment of the Montreal Protocol. Following his undergraduate work, Paul worked as a special guest research chemist for the National Institute of Standards and Technology (NIST) before returning to North Carolina to co-establish the state's first biofuels production facility.

AQILA COULTHURST has been Program Coordinator for STEP since the fall of 2011. Prior to joining STEP, she spent over two years in the production and marketing divisions of the National Academies Press (NAP), the arm of the National Academies that publishes over 200 reports annually. Ms. Coulthurst was involved in several initiatives at NAP including: direct marketing and online outreach; facilitating the sale of intellectual property rights to publishers abroad, and general operational support. Her accomplishments include designing and deploying 100+ emails to over 150,000 subscribers, writing copy for the online catalog and other marketing materials, compiling and analyzing performance metrics to better understand consumer behavior, and developing relationships with agents/publishers in the Asian market resulting in over $40K in intellectual property sales.

Over the years, Ms. Coulthurst has worked in various capacities at Smithsonian Enterprises, the National Community Action Foundation, Kingsley Associates and the Center for Science, Technology and Economic Development at SRI International. She has extensive experience conducting impact assessments and program evaluations. In addition to her interest in U.S. competitiveness and innovation policies, Ms. Coulthurst is interested in how these policies impact development abroad. She spent several years studying U.S. foreign policy and sustainable development at renowned institutions in DC, and while studying abroad in Central America. She has a B.A. in economics and in Spanish, and a certificate in markets and management from Duke University. She also has a Master of Science in Foreign Service from Georgetown University.